'A practical guide on how to start your mobile app company. A must read for anyone who wants to start a mobile app business' – Riccardo Zacconi, Founder and CEO King Digital (maker of Candy Crush Saga)

'The first book to take a detailed, insightful behind-the-scenes look at the mobile app world. A must read for anyone who wants to know what it takes to build an app into a successful business' – Hugo Barra, VP, Xiaomi and former VP Android Product Management, Google

'George Berkowski knows from the front line how to build fast-scaling digital businesses. Don't start one of your own without taking his advice' – David Rowan, Editor, *Wired* magazine

'A fascinating deep dive into the world of billion-dollar apps. Essential reading for anyone trying to build the next must-have app' – Michael Acton Smith, Founder and CEO, Mind Candy (maker of Moshi Monsters)

'I loved this book. It provides an entertaining and rigorous yet very practical guide to success in the brave new world of mobile technology. I will never look at apps the same way again' – Bill Aulet, Managing Director of The Martin Trust Center for MIT Entrepreneurship and author of *Disciplined Entrepreneurship*

'George has combined his own experience at Hailo with a thoughtful study of the iconic mobile companies of this era to provide a helpful guide to entrepreneurs and investors trying to understand the emerging mobile economy' – Adam Valkin, Managing Director at General Catalyst Partners

'Distilling lessons from the leading mobile internet startups, George offers unique insight into the app economy, the biggest and fastest wealth-creating opportunity in history' – Paul Forster, Co-Founder and former CEO, Indeed

'In his new work, George Berkowski delivers a compulsively readable business book; it's a rollicking yet rigorously detailed ride through the process of creating a breakthrough app-based business. Berkowski's engaging account of the rise of the app economy provides an insider's view on entrepreneurial best practices, as illustrated by first-hand insights into the people – and their ventures – who have triumphed by exploiting the mobile revolution. It's a richly rewarding read' – Jeffrey F. Rayport, Faculty, Harvard Business School

HOW TO

PIATKUS

First published in Great Britain in 2014 by Piatkus

A CIP catalogue record for this book
is available from the British Library.

ISBN 978-0-349-40137-9

Typeset in Palatino by M Rules
Printed and bound in Great Britain by
Clays Ltd, St Ives plc

Papers used by Piatkus are from well-managed forests
and other responsible sources.

MIX
Paper from
responsible sources
FSC® C104740

Piatkus
An imprint of
Little, Brown Book Group
100 Victoria Embankment
London EC4Y 0DY

An Hachette UK Company
www.hachette.co.uk

www.piatkus.co.uk

To my parents, and Kasia

Contents

Step 2: The Ten-Million-Dollar App

Achieving Product–Market Fit and Raising Series A Funding

Step 3: The Hundred-Million-Dollar App

Tuning Your Revenue Engine, Growing Users and Raising Series B Funding

Step 4: The Five-Hundred-Million-Dollar App

Scaling Your Business and Raising Series C Funding

Step 5: The Billion-Dollar App
The Promised Land

Acknowledgements

I'd love to thank the following people for all their help, input, opinions and support throughout the process of creating this book: Tina Baker, Niqui Berkowski, Samar Chang, Brad Feld, Jay Bregman, Steve Brumwell, Poppy Hope, Michael Varley, Matthew Osborne, Dominika Dudziuk, Anthony Gell, and my editor Zoe Bohm.

Preface

In early 2011, I had just completed the sale of my startup (a video-dating site called WooMe.com) and was relishing the chance to take a few months off in the sun, when I saw a cryptic post on a tech website from the entrepreneur Jay Bregman and couldn't help but get in touch. I shot him an email, met him in person and found myself intrigued by his idea for a new mobile startup – an app that allowed passengers to hail a taxi from their smartphone. But his approach was fresh and disruptive (and in this book I'll be using 'disruptive' to describe something that brings about a step change, shakes things up a bit). It directly solved the problem of how to build up a community of drivers before any passengers were using the app.

From my very first conversation with Jay I realised that his vision for the company was global – an app that any person could use in any language to catch a taxi in any city in the world. I was sold. Jay needed someone with my experience to translate his great vision into a concrete business strategy, to refine the business model – and build the technology and software to turn it into an app. And so, in early 2011, only a few weeks after the company was incorporated, I joined the tiny startup as the Head of Product. The company was named Hailo.

Since then it's been a meteoric trajectory, growing from a handful of people on the HMS *President* (an old warship transformed into budget office space which bucks, rolls and creaks with the

River Thames traffic) to more than 250 people in 7 countries and hundreds of millions of dollars in taxi fares.

My experience at Hailo taught me a huge amount about what it takes to achieve success on a huge, global scale and proved that anyone with a great idea for an app-centric business has the potential to turn that idea into reality – and a global success. This book is the first step to making that happen.

PART I

Think Big

Chapter 1

The View from the Inside

'Until 1916 we didn't even have any billionaires'
#BILLIONDOLLARAPP

Why This Book Is Different

This book will help you get inside the head of people who have built billion-dollar apps.[1] It will help you to see the world as they see it – and take you on the journeys that they have been through. It will share my insider's view of this world with you – along with the interviews and conversations I have been so lucky to have with these amazing 'mobile' entrepreneurs.

As an entrepreneur – and an engineer – I want to tell you the way it really is. I want to talk about what happens behind the scenes and behind the computer screens. I've read enough stories about the glamorous side of technology startups, where billionaires are created overnight as if by magic – but, to hit billion-dollar heights, there is a huge amount that goes on behind the scenes that your ultimate success will depend on.

I have been lucky enough to work with and meet some of the most talented, passionate – and lucky – entrepreneurs out there. I have also had the chance to work with some of the most experienced mobile-technology investors in the world – including Accel Partners, Union Square Ventures, Atomico, Index Ventures and

Wellington Partners. This book is a distillation of countless late nights, years of hard work and a great adventure.

Whether you're a newcomer to mobile technology, a gifted developer, seasoned entrepreneur or just intrigued by what it takes to build a billion-dollar company in this day and age, this book is for you.

It's not just a theory

My bookshelves are piled high with books brimming with great advice about how to build a great business, about how to cross chasms and be an effective executive. Biographies of Apple cofounder Steve Jobs, investor Warren Buffett, Google cofounder Larry Page, Microsoft cofounder Bill Gates and businesswoman and Facebook chief operating officer Sheryl Sandberg peer down over my desk. But, as I reread these books, I keep finding business strategies that no longer work, or principles that, although only a few years old, seem to be already outdated in the fast-moving world of mobile technology.

Today, the most successful *new* technology businesses are rewriting the rules in real time. A new wave of companies is rocketing to success in mobile technology, and a new type of entrepreneur is driving them towards billion-dollar valuations faster than at any point in history.

I've read numerous accounts of how to build a world class technology company – but no one has yet attempted to package practical, actionable advice about how to build an extraordinary, billion-dollar, *mobile-centric* business into a single book. Over the last few years building up my own mobile startups, I yearned for such a resource, something that would aggregate all the best lessons, and pitfalls, of fast-growth mobile startups. I found nothing. But I did start collecting all the information I could. I asked for introductions to other mobile startups, and I talked to and interviewed countless people. That formed my own personal guide about how to build a mobile company today.

How to Build a Billion-Dollar App is not based on theories. It is not an academic research paper. It isn't a collage of pithy business stories crafted by a journalist. And it is certainly not a guide that promises success and happiness for a four-hour-per-week commitment. It is about reality. It is based on hard data. The advice and information is based on what it took for a select group of entrepreneurs – a group of people not unlike you – with varied backgrounds and experience took to transform their ideas in global, billion-dollar companies in just a few short years.

Combined with my own experiences at Hailo, this book will also capture the very best thinking, experience, insight – and mistakes – from that select group of entrepreneurs who have followed similar billion-dollar paths in the mobile world. My journey building Hailo has given me amazing access to some incredible people and has been indispensible in guiding the decisions that will open the gates to the Billion-Dollar App Club. Now I want to share that information with you.

Inside information

In writing this book, I wanted to draw on advice and insights from only the best companies and the best entrepreneurs. While advice from any source can be useful, the best and most powerful advice is from those people who have actually achieved billion-dollar success in the mobile-app world.

But getting meaningful inside information from entrepreneurs is exceedingly difficult. There are two main reasons for this. The first is that the people building great mobile businesses today are spending every moment of their time doing just that, leaving no time for anything else. The second is that many great entrepreneurs (but certainly not all) are rather coy about sharing too much information about their businesses in case they divulge something that might help their competition gain the upper hand. Personally, I think there is a lot more insight to be gained by actively sharing and soliciting advice than protecting any individual insights I've come across. It's hard enough building an average company, let alone a billion-dollar one.

What I've sought to understand – and distil into this book – is what differentiates the truly *great* companies from run-of-the-mill, *good* companies. I focus on the most spectacularly successful apps – those that have achieved billion-dollar exits, or have undisputed valuations at the billion-dollar level.

If you're an Android or iPhone user then you'll already be an avid user of some or even all of them: WhatsApp (the messaging app), Viber (another messaging app), Square (the payments service), Angry Birds (the now ubiquitous game), Uber (your on-demand chauffeur), Instagram (the social-photography app), Waze (the mapping and social-traffic app), Clash of Clans (the ridiculously popular game from Supercell), Candy Crush (the 'sweet' game from King), Snapchat (the messaging app where your messages disappear after seconds), Flipboard (the app-based social magazine) and Tango (yet another messaging app).

This list is growing all the time, so for the most up-to-date information, visit mybilliondollarapp.com.

While their missions, businesses and stories are very different, they do share a remarkable number of similarities that have propelled them to great success.

In this book I focus primarily on mobile-first companies – ones that have pure mobile DNA. Why? Because being mobile-centric is a different and entirely new way of thinking – and is possibly one of the biggest business opportunities in history.

That being said, there are a number of other billion-dollar technology startups that began as websites or even desktop applications. There's clearly a huge amount to learn from them, not only from the way that they have adapted to the mobile world, but also because they are great examples of modern companies that have grown from ideas to billion-dollar powerhouses by developing better products, great leadership, constant innovation – and, above all, superb execution. Companies such as Google, Facebook, Skype, PayPal, eBay, Amazon, Pandora, Dropbox, Box, Groupon and Evernote fall into this category.

Concrete steps

The world of mobile technology is exciting – and daunting. The mobile landscape is constantly changing: every week seems to herald the arrival of a new mobile device – from smartphones to smart watches, to tablets, to phablets (that's a combination of *ph*one and *t*ablet).

We're also bombarded with new flavours of mobile software – Android's candy-store options include KitKat, Jelly Bean and Froyo. Apple's iOS software is yet different and based on numbers – iOS 5, iOS 6 – and the latest one (at the time of writing, anyway), iOS 7, seems to adopt a rather candy-coloured colour scheme as well. It's all a bit much to keep track of.

No need to worry, though. I'll lead you through all the things you need to know.

Part I starts by exploring what has caused technology to become so mobile so quickly, and why the biggest, most exciting opportunities in commerce, communications and gaming are happening on mobile apps. I'll look at why this shift was inevitable – and the social, cultural and psychological impacts it will have. It will demystify mobile technology so that you can grasp why these changes are happening – and be in a better position to anticipate what the near future will look like.

Part I will also consider what it means – and takes – to think big. Entrepreneurs don't trip over and fall into billion-dollar businesses: they see big problems, *very* big problems that frustrate lots of people, and then create elegant solutions. They do this through a combination of disruptive thinking, solid execution and management of complexity. These solutions are the basis of businesses with huge potential. This section will give you the tools you need to validate whether your idea has billion-dollar potential (and how to adjust it if it falls short).

Part II will guide you through the journey of building a billion-dollar app. There are five key lifecycle steps. The steps map to challenges that need to be met and solved to create a great product, a great team, a great business model and a great company overall. I loosely align these steps with valuations and funding rounds to help you along the journey as, in its broadest sense, the model I've used reflects how venture capitalists have been looking at technology companies for the last decade.

The model I've used is not meant to be perfect – or definitive – because the trajectory of every company will be different. However, it is meant to provide clarity and structure – and distil the key challenges you will meet. There is no one-size-fits-all approach in the world of mobile startups, but, by following, or at least being aware of, the steps mapped out at each stage, you will make sure that all the foundations necessary for building a great, billion-dollar company have been addressed. Let's look at the steps in more detail:

STEP 1: THE MILLION-DOLLAR APP. Now that you've got your head around a billion-dollar idea, how do you get all the basics into place? I'll walk through defining and designing the app you're going to build, finding a cofounder and core team and thinking about how to raise the funding you need to get to the next stage.

STEP 2: THE TEN-MILLION-DOLLAR APP. Your top priority is creating an app that is going to wow people. This is when you focus manically on building the best product for the right audience – something called product–market fit. This is a tough stage, but, once you have created an app that people are happy to pay for, your business will take off.

STEP 3: THE HUNDRED-MILLION-DOLLAR APP. Now that you have product–market fit, it's all about attracting users, and refining your business model to ensure you're a proper, profitable business. It's also time to get a proper company structure in place – including seasoned management – as your business grows.

STEP 4: THE FIVE-HUNDRED-MILLION-DOLLAR APP. Armed with a great app and a reliable way to make money, how do you grow your company quickly and profitably? How do you keep users coming back? And how do you attract users from the four corners of the globe? This stage is all about scaling your business.

STEP 5: THE BILLION-DOLLAR APP. The Promised Land. I'll talk about exits and acquisitions. I'll also talk about how it doesn't get any easier – in fact, now that you're at the top, you're a target. And retaining pole position is a different game entirely. I'll also talk about how every single path to a billion-dollar app is different.

It's a Really Big Number

Before I tell you something about myself, let's just look at this monster number that we call a billion.

One thing that I find difficult to get my head around is how big a billion really is. It is a modern number. Up until the turn of the last century it was left to the domain of astronomers. Until 1916 we didn't even have any billionaires – John D. Rockefeller inaugurated the club on 29 September 1916.[2]

The world population hit 1 billion in 1804[3] – that clearly took a while. We hit 2 billion people in 1927, 3 billion in 1960 and the magical 7 billion in 2011. While it took 123 years to double from 1 to 2 billion, it took only 12 years to go from 6 to 7 billion.[4]

If you think about 1 billion minutes – well that's equal to 1,900 years. If we rewind 1 billion minutes we'd be at the height of the Roman Empire. If we think about 1 billion hours – that's around 114,000 years. At that time we were living in the Stone Age.

Our brains function in the billions: we have 100 billion neurons – or the key cells in the brain. And there are at least 100 trillion (i.e. 100 billion multiplied by 1,000) connections between all those cells – called synapses.

But let's have a look at something a little more relevant to our billion-dollar app. At the beginning of 2014 Facebook had 1.23 billion active monthly users,[5] with an astounding 757 million logging in every day.

Facebook needs between 180,000 and 200,000 servers (the computers that run their website and apps) to support all those users, according to one professional estimate.[6] Think about what this costs to run. Well, according to one of Facebook's financial filings, it invested more than $1 billion so far in server hardware alone.[7]

But, if you think that number is big, wait for it. Google is spending a few billion dollars – every quarter – on supporting the computers it needs to run all its services.[8] In total it has spent over $21 billion on its data centres.[9]

So success is definitely about understanding – and managing – numbers at a billion-dollar scale.

My Story

One of the core themes in this book is that not only is building a billion-dollar app possible, but that opportunity is everywhere and open to everyone, irrespective of their background or experience.

With that in mind, I want to share with you the rather uncommon path that has taken me to this point in my life, and why – above all – I have tried to teach myself to focus on trying to see the opportunity that every situation presents (and having fun along the way).

My journey towards a billion-dollar app was that of a rather normal, at times dysfunctional and at times difficult child. My parents were neither rich nor poor, they didn't have any interest whatsoever in business, and had little grasp of technology (and for years even resisted having a VCR).

I was born in 1977 in Warsaw, Poland and when I was two, my family moved to Australia. I was a rather demanding child, in constant need of feeding. I showed no early signs of ambition or activity (preferring to eat rather than talk, and sleep rather than walk).

When I was seven there was a glimmer of hope. At a kindergarten bake sale I managed to persuade my mother not only to bake scores of cupcakes, but also to buy matching chef attire for my friend and me. The combination of two cute little boys with matching outfits, brimming with energy and big smiles, led to a cupcake sellout. Profits were made, and a salesman was born. My mother is still upset that I haven't bothered to learn to cook.

The summer of 1989 started off wonderfully, but my father had an alternative view and said I had been loafing about for too long. It was only the second week of my vacation and I was about to turn 13. Then he dropped a bomb: 'Son,' he said, 'I bet, even if you *wanted* to, you couldn't find a decent-paying summer job by the end of the day.'

A challenge is a challenge and I accepted. Though I do admit, I was petrified at the idea of getting a job (a wildly alien concept at the time). A full day of going from store to store with nothing but rejections landed me at the greasy counter of a local BBQ chicken restaurant. Less than 15 minutes later I was wearing an ill-fitting, stained apron, wiping down the counters and learning what it meant to earn an hourly wage. That evening I brought home 12 hard-earned dollars and enjoyed my father's incredulous expression. How did I finally land that job? I bypassed my limited CV and opened the conversation with my would-be employer with the tried and tested TV shopping-channel line: try my work right now, and, if you don't like it, then you don't have to pay me.

My high school prom in 1994 taught me a powerful lesson. I wasn't one of the popular or cool kids (my mother insisted on cutting my hair until I was 16), nor one of the sporty types (but I could give you a decent rally on the tennis court). Prom rapidly approached, and I, naturally, was dateless. Then a funny thing happened. The resident hot girl at our tennis club inexplicably started up a conversation with me after one training session. Words were exchanged, a question escaped by mouth, and somehow she accepted my feeble invitation to the prom. On the big day – and I still remember it all too clearly – we made a spectacular entrance at the venue. Mouths dropped. People were perplexed. As it turns out, she desperately wanted to come to our Prom but not a single guy had had the courage to invite her. I was the exception. To this day people comment on that night. From that moment I knew that the potential upside of asking the question you're afraid to ask always outweighs the chance that you'll hear the word 'no' in response.

Somehow, in 1996, I was admitted to the prestigious Massachusetts Institute of Technology (MIT) with the ambition of becoming an astronaut. Despite less than perfect SAT results (SAT is the standardised test for college admissions in the USA), I managed to slip in under the radar. I suspect I was the token Australian – admitted to maintain the broad quota of nationalities on campus (that or they really enjoyed the essay I wrote about my senior-year summer working as an extra on the Aussie television soap *Home and Away*). My four undergraduate years flew by quickly, while slowly, in the background, the Internet bubble continued to expand.

The year 2000 was the pinnacle of tech silliness. Internet start-ups seemed to be a sure-fire path to becoming a millionaire, stock-market launches (or IPOs – initial public offerings) blossomed left, right and centre and investment-banking jobs were being handed out like frozen-yoghurt samples at a shopping centre. I ignored the pleading of my parents to join the Wall Street brigade, and jumped on board the startup train by joining Trilogy, an automotive and telecoms software startup.

Working first at the company's headquarters in Austin, and then the European head office in Paris, I witnessed massive demand for our software. Straight out of college, I was leading teams selling cutting-edge software solutions to companies such as Renault and France Telecom. The company experienced explosive growth – adding around 700 people in 18 months – then, through a combination of a largely inexperienced management team and a massive contract failing to materialise, the company imploded. More than two-thirds of the company was fired overnight. It was an intense – and eye-opening – welcome to the world of startups.

After I had done a stint at a French business school, the spectacular story of MirCorp caught my interest. MirCorp was the company that tried to turn the Russian MIR space station into the first commercial orbital hotel. (In the end the hotel bit didn't work out, but MirCorp did deliver the world's first space tourist – Dennis Tito – to the International Space Station in 2001.) I hunted down the MirCorp founders, and cornered them at a conference in Amsterdam. A couple of weeks later I had a job. I used my tried and tested strategy: try me out for free (I offered to make them the subject of my business-school research thesis) and, if it doesn't work out, don't pay me. I ended up running the company's marketing and partnerships, as well as their online activities, and, with our Russian partners, delivered a wide range of space-tourism experiences to customers all over the world.

In 2005, I decided to get serious about the Internet, and found a job in the e-commerce team at a big telecoms company. I was lucky enough to work for a very talented boss who made it easy for me to excel and assume more responsibility very quickly. I had the chance to build and launch numerous new online products – including advertising solutions, commerce solutions and even various new communications tools. It was exciting to work with a team of designers and developers to put my products in front of millions of users – and get instant feedback. It really whetted my appetite for building my own Internet business.

In early 2007, I was working on advanced, peer-to-peer, video-chat technology in my day job when my boss saw an opportunity: what if we used this super-fast, high-quality communications technology to change the face of the billion-dollar online dating industry? And so I quit with my boss to found WooMe.com – the world's first video speed-dating platform.

Over the course of four years we built our startup between London and Los Angeles with more than 50 people and 10 million users. Early in 2011, Zoosk.com, one of the biggest dating sites in the world at the time, acquired us. It was a wild ride – and a great outcome. All that time we were watching websites go mobile, and had dabbled with creating our own mobile site, but never quite made the leap to a mobile app.

I was keen to take some time off after WooMe was acquired. Just a few days before I set off travelling around southern Europe, I answered a cryptic online post from Jay Bregman, who had just founded Hailo. Jay explained how his approach to creating a single, global app to hail taxis in any city in the world was different from the handful of small competitors already operating. He loved my experience of building a multimillion-user community in the video-dating world – and creating an acclaimed product. Once he'd finished outlining the broad experience of the Hailo founding team I was pretty much sold. And so my journey to delivering my own billion-dollar app began.

Breakfast with three taxi drivers

So how was the idea for Hailo born? Well it all started with breakfast on Charlotte Street in London with three taxi drivers.

Jay Bregman had previously started eCourier – a courier startup armed with advanced algorithms that decreased the time and cost of deliveries, making the entire process much more efficient. While the company was successful, it couldn't displace the massive incumbent courier companies. But Jay still had a burning hunger to build a highly successful technology company.

Jay knew two things all great entrepreneurs know: first, you need to focus on what you know; second, to build something enormous you need to disrupt and reinvent a service that millions of people around the world use on a daily basis.

As eCourier struggled, he focused on figuring out how he might transfer the allocation technology he had developed for eCourier – matching people who wanted to send packages with couriers who could both pick up those packages and deliver them in the most efficient way – to another market. And, most importantly, to a market needing disruption, a good shake-up.

After some research, the taxi industry popped up as an interesting candidate. People use taxis all over the world, in big cities and small cities, and, interestingly, 25 cities in the world spend $1 billion or more on taxi fares annually. On top of that, it turned out that taxi drivers were terribly inefficient – spending up to half their time driving empty cabs, searching for fares.

Jay cofounded the company with Caspar Woolley (the former chief operating officer at eCourier) and Ron Zeghibe, a seasoned executive with experience in private equity and outdoor advertising. Ron had even taken a company public.

And so back to Charlotte Street – and what would turn out to be the genesis of Hailo. Jay, Caspar and Ron were there to meet three entrepreneurial taxi drivers called Terry, Russell and Gary (affectionately known as 'TRG'), who were trying to get their own business, called TaxiLight, up and running. TaxiLight aimed to match drivers coming in and out of London's suburbs at the beginning and end of their shifts, with passengers looking for rides in those directions. The strategy was to discount these longer rides and allow the drivers to earn some incremental income, while passing on a good deal to passengers.

Jay loved the idea, but even more than that he loved the fact that TRG had already signed up 700 London cabbies who wanted to take part in the programme – despite the fact they had no product or app to offer them. It was at this meeting that Jay presented his idea – one bigger and more ambitious than anything the drivers could have possibly imagined – of a truly global taxi-hailing app. The app would launch in London first, and then it would expand globally. Everyone around the table loved it.

Hailo would have six cofounders – three seasoned entrepreneurs and three experienced taxi drivers. The team possessed great collective experience – deep sector expertise, tried and tested technology, years in finance and even a previous working relationship. It was a powerful cofounder mix.

For drivers, by drivers

What became clear from those very early stages was that having the involvement of drivers at the very core of the business was indispensable. As the Head of Product for Hailo, my role was to figure out *what* we needed to build – and *how* to go about building it. I, along with our designer, would work closely with drivers to determine what features should be included in the app, how the pricing should be structured and even the best way to message the drivers. We wanted to create a service that drivers would lo

Hailo's key strategic difference would be its ability to build up a supply of keen drivers well in advance of having any passengers by creating an app that would make drivers' days better and their work more efficient. The app would provide free traffic information, driver-sourced information about where people are requesting cabs, the ability to securely process credit cards via their smartphones with no additional hardware and comprehensive stats and reporting on the fares they were picking up – all for free.

With TRG's input we started work confident that we were taking the right approach. The core of my role was to understand what drivers really wanted – and how to design that into an app they'd love using every day. I also needed to build a product and engineering team that could actually build a robust mobile platform to support thousands of drivers – and millions of passengers. And I had only a few short months in which to do it.

The year 2011 marked another key turning point, before which a mobile-centric solution to the problems of the taxi industry just wasn't possible. The penetration of smartphones was becoming significant – with broad mainstream adoption. Smartphones were no longer just for early adopters: more and more people owned them. This was the very first wave of BYOI – Bring Your Own Infrastructure. From the onset, we knew that, if we had to buy smartphones for taxi drivers, then our profit margins would be a lot lower – and we'd have to get into the business of maintaining all that hardware.

By carefully modelling where smartphone prices were going, we could see that it would be very realistic for Hailo to focus solely on creating the app software – and leave it to drivers to buy their own smartphones. This was a critical decision, as it would allow us to invest more time and money not only developing our software but also focusing more on expanding Hailo to more countries and cities.

Had we instead focused on providing the hardware to drivers (as some of our taxi competitors did) we would have burned through a lot more money more quickly, which would have come at the cost of expanding to additional cities. So it was a decision well made.

Today this decision has resulted in further benefits for Hailo. In most cities – because of Hailo's scale – it is able to negotiate great deals on both smartphones and airtime packages on behalf of its drivers. This has actually become a selling point to drivers thinking of joining Hailo, since they can now get better phones, cheaper, and with better packages.

It was this focus on building an app that drivers loved which would allow Hailo to get incredible initial traction. And it was extending that same strategy to understand people's behaviour that would lead us to design a similarly great app for passengers. But ultimately it was predicting the way smartphones – and the apps on them – would shape daily human behaviour, combined with the massive reach of app store distribution, that would really accelerate Hailo's journey to becoming a billion-dollar app.

Chapter 2

Mobile Genetics

'A typical smartphone user looks at their phone
about 150 times per day'

#BILLIONDOLLARAPP

In early 2014 the number of mobile phones exceeded the number of people in the world today: 7 billion.[1] The number of smartphone users will top 1.75 billion in 2014.[2] In 2013 people spent a whopping $25 billion on apps – an increase of 62 per cent on the previous year.[3]

In the face of these eye-watering statistics, it seems bizarre to think that smartphones didn't even exist in 2007, and it's amazing to think that apps (just a cuter word for 'applications') appeared only a year later in 2008 – launched on Apple's rather revolutionary iPhone.

So why is the mobile sector growing so fast? And why have both the rollout and the usage of the mobile Internet grown so much faster than the desktop Internet? In this section I'll dive into how the last few decades of innovation have converged to deliver history's most powerful computers, in the smallest forms, and at a price that an increasing proportion of the world's population is able to afford. We'll cover everything, from the operating systems used by the two smartphone giants Google and Apple, to the most popular activities on smartphones, to the underlying technologies that make apps so powerful.

Mobile Foundations

Let's get geeky for a moment. Mobile technology didn't become huge overnight. It was a long time in the making – it was just that most of us didn't notice (and frankly didn't care that much). Think about this for a second: the smartphone in your pocket today is about 15 times faster than a Cray-1 supercomputer (1979 vintage), and as powerful as the most powerful computer in the world in 1987.[4]

But it's not *just* about the hardware. A Cray-1 couldn't make phone calls, or tell you where in the world you were standing, or tell which direction was north, or even allow you to read your email. All it offered you was a keyboard and a screen – and the ability to perform millions of calculations per second.

Before computers became mainstream they needed to address a number of issues. They needed to be friendlier to use – the only way you could communicate with them was by a command line interface and complex instruction sets. The advent of the GUI (Graphical User Interface) and its rapid adoption in the 1980s was a huge step forward. The mouse, even though it first appeared in the 1960s, only really gained broader traction in the 1970s and 80s. But the market was still fragmented with IBM, Commodore, Atari and Apple all creating competing hardware, and software. Computer mice, printers and other peripherals didn't play well together.

It was Microsoft who pioneered a huge change in the mid-1980s with the arrival of Windows. Windows was a new operating system – or OS (we'll use that term from now on). An OS is simply a big chunk of software that is the glue of any computer: it manages all the hardware on the computer (such as the keyboard, screen and processor) and manages all the other programs that run on the computer (such as email, the Web browser, your calendar).

Windows exploded in popularity because it was simple to use and provided a visual interface, and banished the command-line interface to the annals of history.

The Windows OS was launched in November 1985 and in just three years it was running on 25 per cent of all computers. In addition to making computers more accessible to users, it also provided a powerful platform for developers. As Windows dominated more and more desktop computers – it ended up running on 96 per cent of the world's computers from 1998 to 2005 – developers could write a software program, and then easily distribute it and have it run on all those computers (no matter who manufactured the actual hardware).

With the Windows OS now managing the complex interface of dealing with any type of computer hardware, software developers were left with the much easier – and interesting – task of building software that could do cool things such as browse the Web, manage your email, generate visual maps and play games.

Over the course of 20 years, the Windows OS allowed hundreds of millions of people to learn to use computers, and millions of developers to create innovative software programs that would generate billions of dollars of value.

Around the early to mid-1990s, mobile phones started to hit their stride. Mobile-phone handsets – along with mobile services – were now affordable, and a mobile infrastructure that enabled people to make calls and send text via SMS was rolling out rapidly around the world.

There was, however, no clear mobile operating system. All the leading handset manufacturers, such as Nokia, Motorola, Sony and Ericsson, had their own proprietary – and closed – systems. As a result there was no simple or effective way for developers to write software that would work on handsets from different manufacturers. There were a few attempts to try to create common platforms – but none managed to get traction.

The opportunity to develop an OS for mobile phones was there. There was an opportunity to dominate software on mobile phones, just as Microsoft's Windows OS had dominated the desktop. But it would be a few years before a winner emerged.

We're going mobile – and there's no turning back

On 29 June 2007, Steve Jobs dropped a bombshell on the world. That day he unveiled one of the most highly anticipated consumer technology devices in history: the iPhone. Over the preceding years Apple had been clawing back its reputation as a true leader in product innovation. The iPod – along with iTunes – had changed the way people discovered, listened to and stored music, displacing CD and MP3 players and providing consumers with a new way to purchase music. The iPhone would continue the disruption: it collapsed the power of a desktop computer into a device that fitted into your hand; it integrated wi-fi and mobile Internet access; and it allowed virtually anyone to write software that would operate on it – all within a beautifully designed handset.

The iPhone delivered such a punch because it combined a number of very significant innovations. First, it packed the power of a desktop computer into a mobile phone. Why was that critical? Because all that computing power would allow it to run a powerful operating system – or OS. That opened the door for Windows-like domination.

Second, Apple introduced the concept of the App Store. It was a centralised place where Apple would manage and distribute all the mobile computer programs (which would be called 'apps') created by developers all around the world. The iPhone was no phone: it was a true mobile computer – and developers were encouraged to create cool apps for it.

Third, the iPhone set a new benchmark for mobile hardware.

Not only did Apple do away with the clunky method of entering letters via a numeric keypad, but they also replaced the pokey BlackBerry keyboard with a virtual one, thus freeing up more space for a larger screen (and more engaging apps). They also filled the iPhone with sensors such as GPS (Global Positioning System, to tell you where you are), a magnetometer (to tell you which direction is north), and accelerometers (to detect whether the iPhone is moving, and, if so, which way).

It was through the combination of all these innovations, launched at the same time, that Apple was able to set the stage for what would be a multibillion-dollar-per-year app industry.

Over at Google, co-CEO Larry Page watched the iPhone launch and realised that mobile was going to become a dominant force. Since 2005 he had been talking to a developer called Andy Rubin about a rather secretive project called Android – a new mobile OS. After only a handful of meetings Page was so impressed by Rubin – and the technology – that he was ready to buy the fledging company and inject massive internal investments into the project. Rubin joined Google in July 2005 and, three years later, Google launched the first smartphone running the Android OS.

Android mirrored Apple's entire ecosystem, but with one key difference – and ultimately a massive one. Android would run on open-source software. That meant that anyone in the world – from an individual developer all the way to a giant smartphone manufacturer – was allowed to use the Android software and tailor it to their own specific purpose. The Android OS could therefore be used, adapted and optimised for use on smartphones produced by competitors such as Samsung and BlackBerry. Apple, on the other hand, had taken a highly controlled, walled-garden approach with iOS (its mobile operating system). Apple would not allow anyone else to use the system, and iOS would run only on Apple-made hardware.

The diagram below shows how dramatically things have changed. When Microsoft launched Windows in the 1980s, it took 12–15 years for it to dominate 96 per cent of all desktop computers.[5]

And yet, when you look to the right of the graph, you can see that iOS and Android have destroyed Microsoft's near-monopoly in a mere five years. People have adopted mobile computing more than three times faster than they did the desktop.[6]

This points in a fascinating direction: people don't want to be artificially stuck behind desktop or even laptop computers. In fact, the data suggests that people want their computers to be as mobile as they are. And herein lies the massive opportunity with mobile – and, more specifically, with mobile apps.

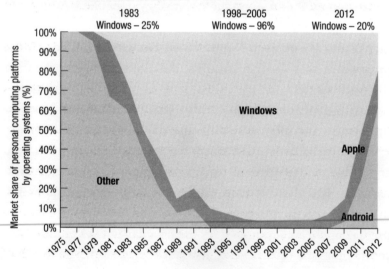

Global Market Share of Personal Computing Platforms by Operating Systems Shipments, 1975–2012

Mobile is winning across the board. China ended 2013 with 618 million Internet users, of which 500 million were mobile Internet users.[7] Eighty-one per cent of Chinese are now accessing the Internet via mobile.[8] Mobile Internet browsing as a percentage of total Internet browsing had reached 23 per cent by January 2014[9] – that's an 83 per cent increase in just a year.

If the above trend continues, then mobile is going to represent the *majority* of all personal computing platforms by 2016.

In 2015, the number of smartphone users is expected to hit 2 billion.[10] The lower cost of mobile devices combined with the growth of mobile broadband networks will actually see the next billion people who come online bypass desktop computing entirely and go straight to mobile. This makes for a huge momentum shift towards mobile.

What are we doing on our smartphones?

It's clear that the demand for more intelligent mobile devices is huge – but what is driving it? One way to understand this accelerating trend is to investigate exactly what we're using our smartphones for.

A typical smartphone user looks at their phone about 150 times per day.[11]

That seems like a ridiculous number, but, when you break it down by activity, it suddenly seems a lot more plausible. Try ruthlessly logging your own smartphone activity for a couple of days. I tried it, and was astonished when I exceeded that number.

All this activity means that we are spending an incredible amount of time interacting with our smartphones. In 2013, the average US consumer spent an average of 2 hours and 38 minutes per day on their smartphone and tablet. That accounts for a whopping 17 per cent of their waking hours[12] – that's almost one-fifth of the time we spend with our eyes open. Wow!

Even more exciting is that those consumers spend 80 per cent of that time (that's right – 2 hours and 7 minutes) using apps and only 20 per cent (31 minutes) on the mobile Web. Apps offer the better mobile experience – and as a result hold four times more of our daily attention than the mobile Web.

So, now that we know that people love their apps, the question begs to be asked: which apps are hoovering up so much of our attention? The diagram below suggests we're a big bunch of time-wasters, spending almost 60 per cent of our time on games, Facebook or entertainment-related apps.

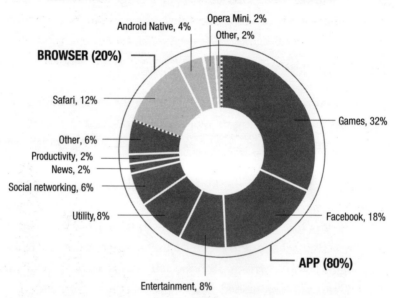

Time spent on iOS and Android connected devices

As you can see from this diagram, there are two apps so signifi-cant in their ability to capture and retain our attention that they deserve a direct mention. The Facebook app galvanises our attention so much that it represents 18 per cent of all time spent by Americans on smartphones. Social is clearly playing an increasing role our lives, especially on mobile. I'll dive into the details later in the chapter.

Interestingly, and this is a reflection of the quality of the app as well, in 2013 Apple's Safari mobile browser captured 12 per cent

of our time, representing more than half of all mobile Web browsing. Given the rather fragmented and competitive mobile browser
space, this is a big achievement which represents another nail
driven into the coffin of desktop browsing.

Generating Billions of Dollars via Smartphones

All this incredible app engagement is already generating billions of dollars in revenue for all the players in the mobile ecosystem. We're going
to investigate the business models that work – both for individual
developers and for the big companies. As the pace of innovation continues to increase, so do the opportunities to generate revenues.

In-App Revenues Are Growing Fast

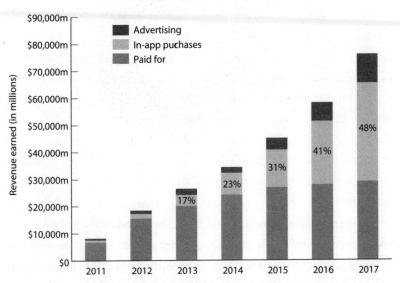

As you can see from the graph above, the first and best-known
revenue stream comes from selling apps through an app store (paid
for). This is big business for Apple and Google, who were the
middlemen for about 90 per cent of the 102 billion apps that were

downloaded in 2013.[13] In 2013, global app-store revenue was $26 billion.[14] Apple and Google keep 30 per cent of these revenues as operators of the stores, meaning that this is the commission that app developers have to pay if they use the inbuilt app-store payment mechanisms. It's worth remembering this figure represents just the revenue from premium downloads and in-app payments; it doesn't include the transactions from apps where you enter your credit card information directly.

App stores are so popular with developers because they have built up massive audiences, and they make it very easy for users to make real monetary purchases with the credit-card details that are attached to their app-store accounts. That means that getting a user to part with real money is as simple as a single click for the developer. The app-store owner manages everything to do with accounts, payments and any hassles associated (such as fraud and charge-backs).

The second major revenue channel is in-app purchases. This has become an increasingly popular topic in the media because numerous apps have seduced children into spending thousands of dollars via apps, unaware that they are doing so.[15] In-app purchasing is a natural evolution of the pay-before-you-download model. Effectively, it makes downloading the app free (thus encouraging more people to download it), and then gives the app developer the opportunity to sell virtual services or products very simply within the app. One of the star billion-dollar apps, the very clever game Clash of Clans, makes 100 per cent of its revenues via in-app payments (Supercell, the maker of Clash of Clans, made $890 million in 2013 using this model[16]). As you can see from the diagram on page 27, in-app payments are projected to be the main source of app-store revenues by 2017.

E-commerce via mobile apps is also a huge channel and is a bit trickier to quantify, as this revenue comes from transactions taking place via apps that go through an app's own payment system – and hence bypassing the inbuilt app-store payment channel used for in-app purchases. The top 500 US mobile retailers – including eBay – turned over \$34.2 billion in mobile retail sales for 2013[17] – up from \$21 billion in 2012. That means that mobile represented about 13 per cent of the \$260 billion total e-retail sales in 2013.[18] Amazon doubled its mobile sales in 2013 to \$8 billion, with eBay pulling in \$8.8 billion for the year as well.[19]

The third major way that apps generate revenue is advertising. While lots of smaller apps rely on this revenue stream, it's very hard to achieve billion-dollar success using this route. But there are two ways. One of our billion-dollar-app role models, Flipboard, successfully executed an advertising strategy via its app magazine to reach its billion-dollar valuation – and has now augmented that model with an e-commerce channel as well. Instagram, the social photography app (purchased by Facebook in 2012), is the perfect example of a highly engaging app that is now rolling out advertising. I'm not going to focus on the intricacies of developing massive advertising revenue for your app at this point. Later, we will explore the risky strategy of building a 'consumer audience' – with the goal of being acquired by a bigger company that can better (and more efficiently) monetise your active users with their own advertising platform.

Apps Make Us Feel Good!

So far, we've seen a clear trend: we're migrating lots of our activities to mobile, using smartphones to communicate more, to play more games, buy more stuff and consume more content. But our spending via mobile has not (at the time of writing) yet caught up to desktop levels – suggesting a huge migration that has only just begun. We are also at a rather fascinating inflexion point: 2013 marked the worst decline in global PC shipments in history, marking seven quarters of decreasing demand.[20] This all results in one very big – and growing – opportunity. In addition to understanding how apps are affecting our lives on a practical level, it's also important to understand the *psychological* and *emotional* effects that smartphones are having on us. In 2013 Facebook completed a study[21] of American smartphone users with some rather fascinating results outlined in the diagram below. Smartphone users were asked, 'How do social and communications activities on smartphones make you feel?' The two strongest sentiments they reported were 'connected' and 'excited'.

Given how personal smartphones are (they are constantly within arm's reach), there is a huge role for apps to play in triggering and maintaining strong emotional relationships. Leading apps capitalise on this by focusing exceptional effort on design, usability, performance and things like the tone of voice used in their copy. Given that people want to be 'excited' and 'connected', it makes sense that anything that is done to maximise those sentiments is only going to make an app more popular.

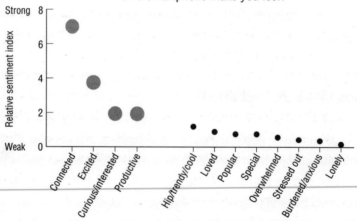

Smartphones Elicit Strong Emotions

How do social and communications activities on a smartphone make you feel?

This strong emotional reaction means that smartphones can have a gripping effect on people. 'Smartphone addiction' numbers are hard to come by, but in South Korea, where smartphone penetration is above 100 per cent (i.e. some people own more than one), the teen smartphone addiction rate is estimated at 18 per cent – that's double the rate of addiction in adults in that country.[22]

A 2013 study by Harris Interactive also pointed out that almost 75 per cent of people are always within five feet of their smartphone, and 10 per cent have used one during sex.[23] The question begs to be asked: is this a sad state of affairs, or a wonderful opportunity? For our purposes we can go with the latter.

What Are Apps, Anyway?

Understanding technology is a crucial advantage, especially in a world where things are changing at an ever-increasing rate. Companies like Apple have done a brilliant job burying technical complexity beneath stunningly elegant and intuitive interfaces, thus delivering value immediately to novice and advanced users alike. After all, Apple pretty much single-handedly created the app ecosystem as we know it today.

This simplicity of interfaces is mirrored on the software development side as well. The barrier to get your own simple app into an app store and into the hands of users is not very complex – opening the world of app development to anyone.

But, in order to build a billion-dollar app, we need to dig deeper into the details. By understanding the very core of what makes apps so powerful, you will possess the insight to develop something *equally* powerful. I'll go through a brief history of how apps came about, and then explore three core reasons that allow apps to deliver a great mobile experience – things you'll need to keep in mind when developing your own billion-dollar app.

A brief history of the app

The term 'app' has been around for a long time. An app (or application) and a software program are the same thing. Thanks to

Apple, though, the word has been adopted by the mobile world, and means either a smartphone app or a tablet app.

This wonderful app ecosystem almost didn't happen at all, according to Walter Isaacson, the biographer of the famed Steve Jobs. When Apple was developing the iPhone, Jobs was initially not too enamoured with the idea that third-party developers should be able to create software to run directly on his beautiful, sleek device – and potentially mess it up. Instead, he wanted developers to create Web apps that could be used through the device's mobile Safari Web browser. Luckily Steve's preference didn't prevail, otherwise the multi-billion-dollar app economy might have evolved very differently.

The challenges with Web apps are numerous, from ensuring compatibility with mobile Web browsers and all their different versions – which involves a lot of testing – to complexities around mobile Web browsers being able to reliably access sensors (such as the microphone, compass or accelerometer) on your phone, and all the way through to how fast the Web app will run if it hasn't been designed to run specifically on your phone. Facebook famously tried to pursue Web apps, only to have CEO Mark Zuckerberg do an about-turn in 2012, stating that it was his 'single biggest mistake',[24] due to those same complexity and performance issues inherent in Web apps.

If it hadn't been for Apple Board member Art Levinson petitioning Jobs, the platform might never have been opened up. 'I called him half a dozen times to lobby for the potential of the apps', but Jobs was against them, says Levinson, 'partly because he felt his team did not have the bandwidth to figure out all the complexities that would be involved in policing third-party app developers.'[25] It was definitely a valid point: Apple does check every single app submitted to the App Store before it's released to the public, which has a certain cost and overheads associated with it.

In July 2008, the App Store was finally launched to the public.

Among the first apps were the *New York Times* news app, BeeJiveIM (one of the first mobile instant messengers), games such as Crash Bandicoot, Rolando and – one that I loved – PhoneSaber, an app that would make a light-sabre noise every time you waved around your phone. Needless to say, a lot of progress has been made since.

Why native apps are best

So even the Silicon Valley elite slipped up on the Web-app-versus-native-app debate – but it's generally accepted that native apps are a better option. Why?

BEST PERFORMANCE. By running natively on a smartphone, an app has the most direct and simple way of communicating to the phone's operating system. That means it renders graphics more efficiently and accesses sensors more reliably and quickly. Web apps – by contrast – must be built to work with the 'intermediate' layer – the mobile Web browser – which adds complexity and inefficiency. To see this for yourself, compare how well Google Maps via a mobile browser works compared with the native app: the native app is faster to load and move and easier to interact with, and generally performs better with complex tasks such as directions.

BEST EXPERIENCE. User experience is key to keeping users happy. Native apps are generally easy to arrange and access on a smartphone (Web apps are clunky to manage). And, depending on the type of mobile Web browser you're using, the performance of the Web app could vary significantly compared with a native app, whether it be the performance of the graphics, the sensors or the touch interactions.

MONETISATION. It is not particularly easy to integrate the various app-store payment systems into Web apps. So, immediately, you're missing out on the powerful payment services that you can build into a native app, enabling users to seamlessly pay from their very first time using your app.

Standing on the Shoulders of Giants

There are two more geeky subjects that you need to be aware of, because they are key factors contributing to the accelerating pace of innovation in mobile.

Open-source software

I've mentioned open source software briefly before, in relation to Android. Open-source software is software that has been made freely available not only to use, but also to change, improve and distribute, and to run your business. On the surface the benefits are clear: it's free to use so therefore you save a ton of money by not paying someone to license it. Just that aspect results in about $60 billion per year of savings to businesses and individuals compared with running proprietary software.[26,27]

That, however, is not the most interesting part. A huge part of the open-source movement is about working on computer code in a public and collaborative fashion. Because anyone can access open-source software, anyone can also improve it. And a lot of people do. As a result, those improvements are pooled, and the quality and performance of the software improves.

This translates into real benefits for app companies, because it means that you no longer have to build your app from a blank page. App developers can go out onto the Internet and find freely available code that allows them to build photo albums, to play music files, to manage user accounts – and, frankly, do a huge amount of tasks in software that previously took a lot of time to write.

By sharing and collaborating in this way, inexperienced developers can build new apps with features that have been refined by numerous developers before them. It means that building an app is many times faster – and more robust.

When you think about the snowball effect this has, it gets very exciting, as it is constantly lowering the barriers to entry into the software world, and at the same time improving the quality of software.

Mash it up now

The second factor doesn't sound too friendly: application programming interface (or API for short). In boring language, this is a common interface that specifies how one computer application can talk to another computer application in a way that can be understood by both sides.

Let's look at a good example. Imagine you wanted to include a map in a new app you are building. You want the app to show the best places to grab a drink in your city. You already have a list of all the great bars you want to include, but you think it would create a good user experience if you could press a button and see the location of the bar on a map. Today, thanks to companies like Google, you can use a mapping API. In a well-documented way, you can have your app talk to Google's mapping service via the API and it will create a lovely map for you, show you the address of the bar on the map, and even show you your current location and directions on how to get there. The whole process takes a few hours to integrate. And – shazam! – it works!

But what if that mapping API didn't exist? Hmm. Well, the only option would be to create your own mapping software. That – actually – is a gargantuan task. So large, in fact, that only a handful of companies have even tried. Google, Apple, Microsoft and even Nokia have all invested billions of dollars in research and development as well as acquisitions and taken years to get their respective services up and running.

And yet this is only one example of communication between apps and other computer applications made possible via APIs.

Billion-dollar apps are increasingly depending on APIs: the Uber and Hailo apps use mapping APIs to show you where your driver is, and payment APIs to securely store your details and charge your credit card; Flipboard uses the Facebook and Twitter APIs so that you can log into their app and share content via those sites with a single tap; gaming apps such as Angry Birds, Clash of Clans and Candy Crush rely 100 per cent on the app-store payment APIs to generate their revenues.

And, naturally, apps can take advantage of APIs in other ways. With Instagram's API you can create a new website or mobile app that automatically grabs your Instagram photos and does cool stuff with them, such as have them printed on demand and delivered to you as postcards, or create a specialised website full of puppy-only photos, or an app that shows Instagram photos by the location they were shot in. The possibilities are limited only by your imagination.

So I hope you can see that understanding the details about the underlying technology can not only save you time and money when developing an app, but can potentially make your app, and the experience it offers users, a lot more creative and attractive.

An Eye for the Future

So far in this chapter we've learned that mobile adoption eclipsed desktop usage by a factor of three, that smartphones are in the hands of a billion users who are spending 20 per cent of their waking hours interacting with them, and that apps are by far the most powerful way we interact with those smartphones.

So, as you play around with your own app ideas, it's important to have your finger on the pulse of what's coming around the corner in mobile. If history is anything to go by, mobile is going to evolve even faster than any of us could imagine, which makes it even more important to watch your blind spot.

The future of mobile is already here. The immediate future is about much better, less intrusive, more intelligent, better-integrated 'interfaces'. The last few decades have laid a robust foundation: smartphones – massively powerful, affordable computers that fit into a pocket – are close to ubiquitous; unimaginably powerful services – from global positioning, to voice recognition, to instant knowledge search – are available for free; and people have invited this technology into the most intimate part of their lives – relationships and financial transactions.

We are finally at a point in history where technology is adapting to the way we live our lives. No longer is a computer a desk-centric destination: it is a mobile companion. With services like the personal-assistant app Google Now, technology is starting to anticipate what we need and want, rather than merely reacting to a request.

As the miniaturisation of technology steadily advances, the age of powerful, helpful, life-changing, wearable computers is upon us. Samsung launched its Galaxy Gear smartwatch in September 2013. Despite dismal reviews – and low sales projections – the company announced it had sold 800,000 devices in the first two months.[28] This is not so much a testament to Samsung's great product design (the reviews are unanimously poor) but it does suggest that consumers have a huge latent desire for a device of this sort.

While the media and blogosphere speculate about a vastly superior 'iWatch' in the offing from Apple – one that will incorporate all kinds of clever non-invasive sensors that may measure all kinds of things, including heart rate, oxygen saturation, perspiration and blood sugar levels in addition to the already commonplace step- and calorie-measuring sensors – other companies are already profiting from wearable technology. The fitness-bracelet market – where devices like Fitbit, Nike's Fuelband and the Jawbone Up lead the market – delivered $2 billion in revenue in 2013. And that number is expected to triple by 2015.[29]

But all those technologies pale in comparison with one. Say hello (or OK) to Google Glass. Google Glass is 63 grams of hardware – a modern-looking set of glass frames (without lenses) sporting a microdisplay that projects an interface (which appears as a floating 27-inch display) into your field of vision. Think of it as an advanced – and heavily miniaturised – version of the Heads Up Display (HUD) systems that fighter pilots use. The device cleverly integrates a video camera that can record videos or photos and an Internet connection, so you can send those images and videos anywhere you like – and it all runs on a version of Google's Android operating system. On top of all that, Glass works mainly via voice controls, so all you have to do is to talk to it; it's science fiction – today. By the end of 2013 over 20,000 people were using it. It is slated for broader public release in 2014.

This is what I meant by interface. What was once a heavy screen confined to the desktop became a smaller screen able to be carried in your pocket, which will now become a screen and voice-control system so light – 63 grams – that you barely notice you're wearing or carrying it. And it gets spookier. There are projects under way that will enable features like Google Glass to be packed into a contact lens.[30] The implications of such an unobtrusive – and powerful – interface are simply jaw-dropping.

If we go back to the beginning of the last technology cycle – that of the smartphone, kicked off by the iPhone in 2007 – we can see how quickly a touchscreen interface, a powerful operating system, integrated sensors and a ubiquitous mobile Internet connection changed our lives. It was just a matter of years. When the next technology cycle begins – and it will undoubtedly be something more wearable – it will begin with huge swathes of the ecosystem already in place. The time to get 1 billion active users of a gizmo like Google Glass will be a lot shorter than the eight years it took the smartphone to smash that milestone.

While Google Glass has received a lot of attention because of Google's profile, another equally fascinating, and potentially even more disruptive, technology company has captured headline. It is called Oculus VR and it might just be the first company to bring virtual reality to the masses.

The company's founder Palmer Luckey is a self-proclaimed virtual reality enthusiast and hardware geek. He launched a campaign on crowd-funding website kickstarter back in 2012 to build the Oculus Rift – a groundbreaking virtual reality headset for immersive gaming. The campaign was beyond successful and raised not only $2.4 million in funding, but also won the support of three huge gaming companies: Valve, Epic Games and Unity. That success attracted some of the gaming world's best talent, almost $100 million in venture capital funding and the acquisition of the company by Facebook in March 2014 for $2 billion. While virtual reality headsets are very desktop-centric today, the experience they could deliver in a mobile environment, combined with the pace of miniaturisation, might see them become part of the mobile app ecosystem more rapidly than anybody might expect.

The more you know about what is coming, the better you can make sure your app idea is perfectly placed to take advantage of what's next.

Chapter 3

A Billion-Dollar Idea

One big reason I wanted to write this book is to encourage people to think big. But I wanted to help you to think big in a structured, realistic way. We can all certainly imagine how we'd *spend* a billion dollars, but it's a lot harder to put a plan together about how to realistically *generate* a billion dollars. There aren't any guides out there about how to think at a big – a truly huge – scale from the beginning. I'm going to change that.

No matter what happens in life, you probably won't hit your precise goal, but the higher you aim, the higher you are likely to hit. So it makes complete sense to start with the biggest possible vision, so that you stretch yourself from the very onset. Why not shoot for a billion-dollar idea, and then have reality thrust you back to a $500 million one?

'You will only have one great idea in your life – make it count!' wrote Eric Jackson in a *Forbes* article.[1] While I'm not sure if Jackson has ever run a company, he is right: truly great ideas don't come around that often. So, when your number is called, make sure that you're *so* well equipped and prepared that you can really belt it out of the park. That's the goal of this book – to prepare you for your chance to build a billion-dollar app.

Start with Big Problems

> Any big idea is going to take a while to get there. By definition, if it's big, and no one has done it before, it's not going to be 1-2-3, 'We got it!' There is going to be a dark period in there, because you don't know what the key to getting there is. You have to be willing to be in some murky territory, and be prepared to invest, if you really want to do something different.[2]
>
> *– Evan Williams, founder of Blogger, Twitter and Medium*

That's pretty good advice from someone who invented a new format of human communications. During its early days, Twitter was massively criticised because people didn't understand it. It's a great example of the intersection of perseverance and luck – but it's not the best mobile-first example.

Let's dig into the ideas that have borne real mobile-first, billion-dollar apps, and then let's try to build that into a framework from which similar quality ideas can be built.

Jack Dorsey, the cofounder of Twitter and friend of Evan Williams, explains the story of how his mobile app Square came to be. It starts with a friend – glassblower Jim McKelvey – who had a customer interested in buying one of his pieces for $2,000. The buyer then asked whether he accepted credit cards. Unfortunately, Jim didn't. And as a result he lost the sale. In Dorsey's mind this generated a huge question: why couldn't anyone with a smartphone become a card-processing merchant? After all, smartphones had way more processing power than typical credit-card terminals;

they also had superior screens and already had mobile Internet connections. The only component missing was the actual card reader. And so the idea for the Square app – along with its card-reader attachment – was born. We'll investigate Square further later on in the book.

Jan Koum was a ten-year Yahoo! veteran. In that time he developed a deep distrust of how ads corrupt the relationship between a company and its users. He believed that advertising was so invasive – and disliked by consumers – that he wanted to build an app that would allow anyone in the world to SMS for close to nothing. Koum wanted to shun advertising entirely, and focus exclusively on utility and user experience. And so he built a simple – but powerful – messaging app called WhatsApp, initially asking users to pay a one-time charge of $0.99 for unlimited text messages (it subsequently changed its pricing and charged that amount per year, as we'll see later). Within five short years, in early 2014, WhatsApp had 450 million users across just about every country in the world. It was clearly solving a big problem in way that users loved. Facebook acquired them for a brain-melting $19 billion in February 2014.

Niklas Hed didn't so much address a big problem as fill a big hole. When the iPhone arrived and started changing our mobile lives, gaming companies took time to adapt. No one was focused on developing a game tailored for the touch interface of the iPhone. Sure, there were successes such as Tap Tap Revolution (co-founded by Andrew Lacy, a fellow Australian and friend), but the market was lacking something truly novel, international and just plain great. Angry Birds was very specifically designed to scratch that itch. It took advantage of the iPhone as a brand-new gaming platform and became a global phenomenon.

Mike McCue and Evan Doll wanted to make content on the Internet more beautiful; they wanted to consume it more as one would a beautiful, visual magazine. When Apple launched the iPad in early 2010, McCue and Doll knew that this was the platform to realise their vision. Everything snapped into place, and they rushed to build a gorgeous social magazine platform: Flipboard. Named iPad App of the Year later in 2010, the app is flipping some 10 billion pages per month, generating big advertising revenues and was valued at $800 million at the end of 2013.

Noam Bardin, Ehud Shabtai, Amir Shinar and Samuel Keret were all sick of sitting in traffic. They wanted a real-time, reliable way to report and share traffic information when they were stuck in the middle of it. They created a clever app that allowed users to do just that. Within a few years their app, Waze, amassed 30 million super-engaged users around the world, and developed its own unique mapping technology. Despite the fact that they were making zero revenues, Google acquired the company for a cool $1.1 billion in 2013.

Privacy and anonymity have more or less gone out of the window with our increasingly online lives. Evan Spiegel and Bobby Murphy created an app whereby you could share annotated photos with your friends – but, once read, the messages would disappear for ever. The app is called Snapchat. A lot of people dismissed the idea – but then something happened. Teenagers found the idea brilliant. Suddenly they could be themselves, share whatever photos they wanted, and rest assured that each photo would be seen once and then disappear in 10 seconds. In just a couple of years the app attracted millions of users sending hundreds of millions of snaps every day. It's become so popular that Snapchat turned down acquisition offers of $3 billion from Facebook and Google, according to the *Wall Street Journal*.[3] We'll talk more about Snapchat later.

All these are great examples of apps that tackled big problems felt by millions of users. They all pinpointed a real need – and then went ahead to address it. While each app started with a single user, they all built large, engaged user bases by focusing on one major issue.

What People Love, What People Need

One approach to coming up with a big idea is to understand what people love to do, and what people need to do. It's true that this varies wildly by geography, personality type and a myriad other factors. But, according to Donald Brown, a professor at the University of California, there is actually a common denominator to all human civilisations – a certain set of 'attributes' – which makes us fundamentally human. Brown has termed these the 'human universals'.[4] Let's use this as a starting point.

According to Brown, the human universals 'comprise those features of culture, society, language, behaviour and psyche for which there are no exception. For those elements, patterns, traits, and institutions that are common to all human cultures worldwide.'

There are 67 universals in the list that are unique to humans: age grading, athletic sports, bodily adornment, calendar, cleanliness training, community organisation, cooking, cooperative labour, cosmology (study of the universe), courtship, dancing, decorative art, divination (predicting the future), division of labour, dream interpretation, education, eschatology (what happens at the end of the world), ethics, ethno-botany (the relationship between humans and plants), etiquette, faith healing, family feasting, fire making, folklore, food taboos, funeral rites, games, gestures, gift giving, government, greetings, hailing taxis,* hairstyles, hospitality, housing, hygiene, incest taboos, inheritance rules, joking, kin groups, kinship nomenclature (the system of categorising relatives), language, law, luck superstitions, magic, marriage, mealtimes, medicine, obstetrics, pregnancy usages (childbirth rituals), penal sanctions (punishment of crimes), personal names, population policy, postnatal care, property rights, propitiation of supernatural beings,

---

*I added this one myself, just to see if you were paying attention.

puberty customs, religious ritual, residence rules, sexual restrictions, soul concepts, status differentiation, surgery, tool making, trade, visiting, weather control, weaving.

My point here is that if your idea resonates with a human universal, you will maximise the universal appeal of your app. Solving a 'universal' problem creates a much bigger market opportunity than solving a geographically specific, language-related or generally niche issue not shared by a huge number of people.

On the flipside, not every human universal maps to a billion-dollar idea. But the list of universals does provide a great checklist, so it's worth checking to see if you can match apps that correspond to each one.

When I was doing this exercise, I came across a fascinating example. I discovered a free app that, despite having more than 129 million downloads[5] and massive daily usage numbers, has garnered very little media attention. It is called YouVersion.[6] It's a free Bible app that offers 600 translations of the Bible in 400 languages. It's a billion-dollar opportunity that maps directly to the 'religious ritual' universal. It doesn't earn much revenue today, but that just may be a matter of time.

A Hundred and Fifty Times a Day

We should thank Tomi Ahonen for what he's done. He publishes an annual *Mobile Almanac*, and one of the most interesting things he uncovers is precisely what each one of us is doing on our smartphones. According to his research, the average person checks their mobile phone an astounding 150 times per day.[7]

Let's have a look at what those interactions and activities are, and let's see where the potential billion-dollar ideas exist, or are still waiting to be uncovered.

MESSAGING-RELATED, 23 TIMES PER DAY: Messaging is a fiercely competitive space. Everyone from Facebook to Google has a messaging app. But, as a number of players have shown, it is an area ripe for disruption. WhatsApp, Tango and Viber have both built billion-dollar propositions here. Asian apps – such as Line and WeChat – have created massive multi-hundred-million user bases here. And in just a couple of years Snapchat turned messaging on its head – and turned down billion-dollar buyout offers – by making messages more interactive (you can scribble and comment on your photos and then send them to friends and, as we saw earlier, they disappear after 10 seconds – all for free).

The takeaway here is that, if the app offers an experience that hits a nerve – a latent psychological or behavioural need (I want to be anonymous with my messages) – then it explodes. If Snapchat can disrupt the market, then clearly so can others.

VOICE-CALL-RELATED, 22 TIMES PER DAY: Mobile carriers still carry the vast majority of calls over non-data networks, but plenty of apps have come to eat more of their pie. Skype is a leader in voice calls with its mobile app (200 million active users and $200 million in annual revenue), as is the Viber app, which amassed 300 million active users by early 2014.[8] Like Skype, it uses instant messaging and a voice-over-Internet protocol (VOIP). Google has its own Hangouts app, and Apple has its Facetime app built right into the iOS platform. As mobile carriers realise that their future lies in data, they have offered unlimited national calling packages, removing revenue opportunities for apps that want to compete. The majority of the opportunity here is in international voice calls, but that domain has been rich with competition for decades, so the opportunity is no longer clear-cut.

What is clear, though, is that voice calls are becoming absorbed into the broader messaging category, with people sending SMSs, chat messages, photos, MMSs, video and audio calls, and even recorded voice messages, interchangeably.

CLOCK, 18 TIMES PER DAY: No one has yet innovated on the basic clock in a way that makes money in any serious way. What would make the killer clock app? There is clearly a captive, global audience here.

MUSIC PLAYER, 13 TIMES PER DAY: This is a tricky area and there are many opportunities for innovation beyond a straightforward music player. Music lives as a larger ecosystem – from the discovery of good music to the purchase, to the playback and organisation. It's a very fun space and attracts a lot of attention, but it's also very competitive for that precise reason. iTunes, Spotify and Pandora have all become billion-dollar leaders in the space, but not without long, gruelling stories to tell.

Is there a big, new, fresh opportunity here? Probably. But having to deal with record labels and international licensing agreements is going to be a challenge.

GAMING, 12 TIMES PER DAY: Gaming has been one of the most responsive and successful sectors on mobile. It has a long history of working with new and mobile platforms, from consoles to the Nintendo Game Boy, the Sega Game Gear and the Sony PSP, so it was no surprise when the industry jumped on board iPhone so quickly. Angry Birds, Candy Crush, Clash of Clans, and Puzzles and Dragons are all billion-dollar-app franchises. Supercell, the maker of two of these hit games – Clash of Clans and Hay Day – made $892 million in revenue in 2013 and profits of $464 million.[9]

There is undoubtedly huge opportunity in app games but, with an increasingly competitive marketplace and gamers with higher standards, it's still going to be tough to create the new billion-dollar hit.

SOCIAL MEDIA, 9 TIMES PER DAY: Currently, this space is dominated by the likes of Facebook, Twitter and newer players such as Pinterest. There are also huge players in Asia, with the likes of the Chinese microblogging site Weibo owning their own markets. Given that the dominant social media have been increasingly integrated into the operating systems at a native level, displacing these guys will be hard. But as new players keep springing up – and blurring the lines between messaging and social media – there is a good chance this will be disrupted.

ALARM, 8 TIMES PER DAY: This is a big opportunity. There are lots of alarms apps in the market – and it's clear we all use at least one of them on a daily basis. One innovation that has appeared in this sector is sleep-related alarm apps, such as Sleep Cycle, which attempt to wake you at the most optimal moment. But no single player seems to be dominating the market at the time of writing, and there is no single must-have app (partially because Apple's native app is pretty good). Personally, I think there will be a big winner in this space. It won't be anything obvious and I suspect it will be a very clever mashup app that takes us all by surprise.

CAMERA, 8 TIMES PER DAY: As of early 2014, we're sharing more than 500 million photos a day.[10] That is on track to double what we shared in 2012 – and that growth is accelerating. Taking photos is universal, and, with more than 5,000 camera-related apps in the App Store alone, the competition is fierce. From the built-in Camera app on iPhone to Camera+ to Instagram, it's a bit too easy to make a camera app – but this is a huge and fast-changing market. If I were a betting man, I'd bet there's another billion-dollar app here.

NEWS AND ALERTS, 6 TIMES PER DAY: This is another great – and massively fragmented – sector, although you could argue that

Flipboard is breaking into this space and has already built a billion-dollar head start on the competition. And, strictly speaking, it's not just news or alerts, since it is taking a broader magazine approach. There's also already a mass of other news-reader and news aggregation apps, but none have caught the attention of the public as much as Flipboard.

CALENDAR, 5 TIMES PER DAY: I believe someone will launch a billion-dollar calendar app in the near future. Calendars are inherently social, since we do things with colleagues, friends and family all the time. Calendars outline where we want to be at given times, and we expect, and want, them to prompt us with alerts. It's an indispensable app that attracts a lot of attention and is crying out for disruption!

SEARCH, 3 TIMES PER DAY: This refers to actually searching the contents on your smartphone rather than searching the Internet. Trying to compete with a native function that is built into the OS is a big battle and there's no clear problem to solve here, since the existing solutions are great. And, while I'd love someone to challenge Google at its core business, well, let's say that's a pretty tough ask for a new mobile startup!

OTHER RANDOM WEB BROWSING, 3 TIMES PER DAY: It's clear from the data that smartphones are currently highly app-centric for a combination of reasons, ranging from apps' ability to deliver individual tasks better than a Web browser could to the simplicity and clarity of having a single act best execute a single task. The big question is how this will evolve, and how the next generation of mobile browsers will challenge the current, rather simplistic, app architecture.

CHARGING PHONE, 3 TIMES PER DAY: The market for charging solutions and devices has exploded with smartphones. Whether they be simple battery packs or multi-device chargers, solar chargers or even the very cool inductive wireless charging pads (which have yet to catch on in any significant way), there is a clear demand by users to keep their devices full of juice. Given that this is outside the realm of apps (for the near future), I'm not going to focus on it too much.

VOICEMAIL, 1 TIME PER DAY: People don't like checking voicemail – I don't know anyone who does. As behaviours have changed – triggered by the prevalence and convenience of numerous messaging apps – the relevance of voicemail has diminished. Apple delivered a serious improvement to the voicemail experience with 'visual voicemail' (a visual interface that downloads and allows local access to your voicemail messages), whereby an audio file of your message is conveniently downloaded and can be accessed with a single click. Once again, that is integrated into the OS, and took significant effort on Apple's part to make it a reality (think negotiations with every individual mobile network operator). The last serious attempt at a voicemail startup was Spinvox. Despite blowing through $100 million of venture capital,[11] everything ended in tears when it was discovered that its speech-to-text technology was little more than overworked call-centre employees transcribing the audio messages themselves.[12,13]

Voicemail as we know it is probably not a great opportunity, but the broader messaging arena is where the action is proving to be.

OTHER MISCELLANEOUS USES, 10 TIMES PER DAY: This catch-all category represents a good 6 per cent of all interactions with mobile phones. So, while the above opportunities are clear and have attracted lots of competitors, there is still big opportunities around not-yet-invented apps, which have the chance to capture our attention in new and novel ways.

Sharing Big

As humans, we love to share. It is part of our nature; it's been ingrained in our psychological makeup. On a practical level, sharing makes sense because we often need the support, skills and insights from others to help us through our day, our jobs and our lives in general. But sharing also just feels good because of the way our brains are wired. According to Harvard University professors Diana Tamir and Jason Mitchell, sharing information about ourselves is intrinsically rewarding and gives us a few squirts of dopamine every time we do it.[14]

When coming up with your big idea or big problem to solve, think about whether it is inherently social, or whether it could be *made* social, thus rendering it a lot more disruptive and a lot more powerful.

People love to share rich content – such as photos, news and magazine articles – and this builds very strong network effects. A great example is Instagram. One of the main ways it drove growth from the very first day was by allowing users to simultaneously share their photos on Facebook and Twitter from the moment the photo was taken using the Instagram app. This massively increased the reach of the new app to big social networks – with very compelling photo content. This promotion of the Instagram app on Twitter and Facebook led people to its website to download the app. The success of Flipboard – the social magazine app (see earlier) – is completely off the back of this premise as well.

What is disruptive, however, is taking something that was not social and shareable – and reinventing it completely. Let's take the example of Groupon. When it launched, its premise was simple: if you want to buy a great product at this 50–75 per cent discounted price, you need to get 100 other people to buy it too within a certain timeframe. Suddenly, Groupon created a very selfish – and social – incentive for people to tell others about their sales. Today, more than half of Groupon's revenues come from its mobile app. And the social component is going strong.

Let's dig a bit deeper into how much people share. The chart below gives us a great idea of how much different countries like to share. The results are fascinating. Up-and-coming markets such as Asia and the Middle East are sharing a lot more than the UK and the US (11 per cent and 15 per cent respectively) and compared with a global average of 24 per cent.[15]

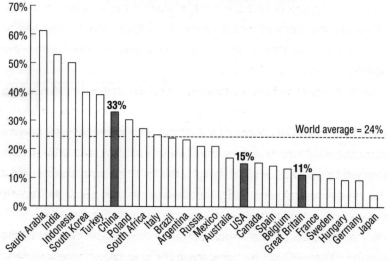

Percentage of Respondents Indicating They Share 'Everything' or 'Most Things' Online

In China about 33 per cent of the population is sharing *just about everything* online. You can see why we're seeing massive growth and interest. This is more than double the rate of sharing in the United States. Why is this important? It means that big ideas have the ability to spread and take off a lot faster in some parts of the world – and that means very lucrative windows of opportunity.

Disruption Comes in Many Flavours

So we've already had a lot to think about in terms of getting to a big idea. And, unremarkably, size is not everything that matters. To cut through the noise and capture attention, it's about being fresh, innovative or, better yet, disruptive. It's not trying to create something slightly better, with marginal improvements (current apps are well positioned to do that – and will). The key is to have a disruptive idea.

A disruptive idea is one that delivers a step change – an idea that is hard for an existing player to copy. This is the wow factor. Wow is going to play a big part in this book, because an idea, a product that wows its audience, is one that can rise to the top and truly stand out. People talk about wow, and people keep using something that wows them.

Rethinking how to solve existing problems is how people disrupt. Our stellar apps have been able to disrupt their own markets – and then create their own perfect storms. The press love to paint a picture of serendipity, with stories of an average person who was 'in the right place at the right time' and how 'you too could build a billion-dollar app with $1,000 and no programming skills'. These superficial, romantic treatments rarely capture the true story.

The best disruptions appear simple – they are best because they are the simplest to communicate and the simplest to understand by the largest number of people. Mass appeal is a core component of far-reaching disruption. Unsurprisingly, the apps with indisputable billion-dollar status embrace simple propositions. Despite the veneer, simple ideas are rarely simple to execute.

Great, disruptive entrepreneurs need to understand the capabilities of the technology available to them, the necessity of building new platforms, how to integrate virality into their products and, perhaps most importantly, the power of timing.

Don't get me wrong: there is *almost always* an element of luck involved (and often significant opportunity cost). But being an entrepreneur is not for the conservative. Nicholas Nassim Taleb (author of *The Black Swan*) would question the viability of betting on low-probability, high-impact events, what he calls black-swan events, but that is the business of entrepreneurs: manufacturing opportunities that are rare and complex and ultimately yield huge returns.

So let's take a deep dive into the key disruptions delivered by our billion-dollar apps, and expose the critical factors that you need to take into account.

Why hating advertising pays more

Jan Koum and Brian Acton both hate advertising. Both were long-time Yahoo! employees who, in 2009, amid the aftermath of the financial crisis, founded a messaging service that aimed to be the biggest cross-platform one in the world. Three years later, they were bigger than Twitter; just five years later Facebook acquired the app for a staggering $19 billion.[16] Jan Koum, WhatsApp CEO, grew the app from 200 million[17] active monthly users to more than 450 million[18] during 2013. That's the fastest growing company in history; in comparison, after four years Facebook had 145 million users.[19]

WhatsApp is a cross-platform smartphone app that lets you send text and picture messages for free, using your data allowance. There are numerous apps and services like this – but WhatsApp is the biggest. As of January 2014, WhatsApp was processing 50 billion messages every single day.[20] That number is more than all the SMS messages sent around the world on a daily basis[21] – it's incredible.

What differentiates WhatsApp from the other players is its business model. It does not rely on advertising, nor does it force you to buy virtual goods. Initially it focused on iPhone users and charged a mere $0.99 per download for unlimited use, but the company moved to a $0.99 per-year model in late 2013. When it launched the app on Android, it discovered that those users were not willing to stomach the $0.99 price (more on this later), so it made the app available for free for Android. Even so, the company has hundreds of millions of dollars of recurring revenue still coming in.

After ten years working at Yahoo!, Koum developed a deep distrust of how advertising can corrupt the relationship between a company and its users. 'The user experience would always lose, because you always had to provide a service to the advertiser.' That's even more acute on mobile. 'Cellphones are so personal and private to you that putting an advertisement there is not a good experience,' he said. Given that we're using these devices 150 times a day, and given the app's great success, this argument is clearly valid. And, sure enough, this was a rather amusing discovery on the company's blog, when it quoted the Brad Pitt character Tyler Durden, from the movie *Fight Club*, as saying:

> 'Advertising has us chasing cars and clothes, working jobs we hate so we can buy shit we don't need.'

People want a messaging app that is simple, fast and useful. They also want all their friends to be using it. WhatsApp delivers on all those fronts. In an interview with the *Wall Street Journal*, it is poignant to see Koum focused more on talking about optimising server code to ensure messages are being moved more quickly, more efficiently and more reliably than other hot topics (such as valuations and buyouts). It is impressive that the company still has only around 50 employees in 2014. That is a testament to its clinical focus, and to the calibre of the people it is hiring.

Koum points out that WhatsApp has become a key sales proposition for mobile carriers in some countries: people come into stores asking for a new phone, specifically because they want to use WhatsApp. In some markets carriers have even created specific packages that allow users to roam and still use WhatsApp – sending unlimited messages and pictures via the app for just $5.

WhatsApp was acquired by Facebook in February 2014 after a fierce bidding war with Google. The $19-billion-price tag represented almost 10 per cent of the market capitalisation of Facebook at the time of the acquisition. Jan Koum's vision is to put the app on every single smartphone in the world[22] and this goal is definitely one of the reasons that it was such an attractive acquisition for Facebook. WhatsApp is a simple – but very disruptive – company that generates hundreds of millions in annual revenues. It raised only a small amount of funding (for strategic reasons) and has been profitable from the very beginning. As a result it has been able to write its own destiny.

Anonymity is worth a lot

This is a story about Evan and Bobby. They were students at Stanford University. In April 2011, Evan Spiegel presented an idea for his product-design class: an app 'where friends could share photos that would disappear – forever – in a matter of seconds'. That app, as we saw earlier, was called Snapchat.

It was launched in September 2011 in Evan's dad's living room.[23] 'Everyone said, "That is a terrible idea,"' Spiegel remembers. 'Not only is nobody going to use it, they said, but the only people who do will use it for sexting.'

By 28 November 2012, users had shared more than a billion photos on Snapchat. By December 2012 it was being used 30 million times a day, with users sending more than 20 million photos per day. In late 2013 this number reached 350 million per day.[24]

The Internet has changed. Spontaneity is now a punishable offence. Instead, 'People are living with this massive burden of managing a digital version of themselves,' says Spiegel. 'It's taken all of the fun out of communicating. The main reason that people use Snapchat is that the content is so much better. It's funny to see your friend when they just woke up in the morning.'

And that is how Snapchat disrupted communications. It has made an entire generation – a much younger one – feel liberated again. While the mainstream media continue to struggle with this concept, the app attracts masses of users – and keeps driving eye-popping numbers month after month.

You could argue that Snapchat is just a feature. But, goddamn, it's a great feature! It's amusing that by June 2013 this company was valued at over $800 million[25] – but that's because at the time it had 5 million users sending more than 200 million snaps[26] (messages), photos and videos every single day – that's up 25 per cent from 150 million announced by their CEO in April 2013.[27] That's ridiculous growth. More recently, they turned down acquisition offers of $3 billion and $4 billion from Facebook and Google respectively.

On average, a user uses the Snapchat app 34 times a month. That means by the end of 2013, Snapchat's 350 million daily snap number matched the number of photos users uploaded to Facebook.[28]

In the true spirit of a lean startup, at the end of 2013 the team was still only around 40 people. Snapchat is still a story in progress. But the lessons are clear: it focused on a universal need, messaging, mixed it up with a true innovation, anonymity, and then focused on a great experience and performance.

Designed to be touched

Angry Birds is one of the most popular games – and brands – in history. It rocketed to billions of downloads, and similar revenues, because of two things: design and touch.

The game was the product of Finnish design studio Rovio. The team had been working in gaming for the better part of a decade, making games for other companies, trying their own titles as well, but never making it big.

When the iPhone came along their eyes lit up – here was a new platform. That meant a new opportunity. They dreamed of designing a fresh, new type of game that everyone in the world could play. They wanted to exploit the opportunities offered by the big, touch-enabled screen of the iPhone by focusing their user interface on visual, audio and touch elements. They wanted to seduce users with interaction and animation – not with words or text. They wanted to create a game that would resonate with basic human psychology. They took a gamble and wrote Angry Birds specifically for this new amazing mobile platform.

A shiny white square

Mobile also introduces opportunities for hardware disruption. Smartphones have spawned not only a world of apps, but also an entire ecosystem of accessories, plugins and add-ons. One of the most amazing was created by Jack Dorsey, a cofounder of Twitter, and now the CEO of Square, the payments service.

Square, as we saw earlier, was born out of frustration. Given that smartphones have more computing power than any cash register or credit-card-processing machine, surely they could become a device to accept credit cards?

Dorsey was a seasoned entrepreneur and found the right people to help build a hardware prototype. White, shiny and in the shape of a thumb-sized square, the Square accessory plugs into the headphone jack of (most, but not all) smartphones. The user can then swipe any credit card through their device, and – hey presto! – their credit card has been debited.

Square tackled a big problem, enabling anyone to simply, easily and cost-effectively, become a credit-card-accepting merchant. By the end of 2013, they had empowered more than 2 million small businesses to accept credit-card payments, and had processed over $15 billion in transactions.

Summary

Things are moving very quickly – and it's clear that they're moving in the direction of mobile. As human beings we prefer to be mobile. Advances in hardware, sensors, batteries, operating systems and platforms are now adapting to our lives and the way we prefer to do things. This is a huge shift towards convenience, usability and utility – and it's a shift that will only become more pronounced.

At the same time we've seen that you need to focus on big universal problems or needs – combined with a disruptive approach – to kick-start a billion-dollar app. As we'll discover in Part II, founders of groundbreaking apps don't just stumble into something great: they have fantastically ambitious visions from Day One. It is a combination of this vision to solve existing problems in novel ways, the refusal to take no for an answer and persevering in the face of scepticism that has launched apps that have changed our lives.

You're now in a great place to understand what the characteristics of a disruptive, billion-dollar idea are. You've seen some ideas about how you can develop your own. And, most importantly, we've seen why the best chance of achieving success is riding the mobile wave.

PART II
The Journey

PART II

Chapter 4

It's Bloody Hard

'0.07 per cent of funded startups become billion-dollar companies'
#BILLIONDOLLARAPP

L et's start with some depressing news. It's pretty obvious that the process of building a successful technology startup – let alone an app – is terribly difficult. But how tricky is it? It's worth looking at the numbers.

From 2004 to 2014 at least 43 companies have hit a billion-dollar or greater valuation (either by venture capital firms or the public stock market) with some experts putting the number closer to 70.[1,2] This number will always be approximate, since private companies don't really share their valuations publicly. It's pretty tough to estimate how many companies have actually been funded – again, because the information is inherently private – but, if you estimate that about 6,000 tech companies are funded in any one year, you get a number of about 60,000 from 2004 to 2014.[3]

So that generates a rather interesting number: 0.07 per cent of funded startups become billion-dollar companies. In other words, you have a 1-in-1,538 chance of reaching the Billion-Dollar App Club, once you have secured professional investment.

There are of course many other scenarios – other than hitting the billion-dollar mark – that constitute success. And, once you have broken through the tough early stages, the numbers definitely shine a positive light on things, with the likelihood of success increasing throughout the lifespan of your startup.

The Rest of Us

What happens to the other 99.3 per cent of startups, though?

According to the database Dow Jones VentureSource, about 11 per cent were acquired or went through an IPO. If we focus on the data about US startups (which is most reliable) we see that the average successful startup that exited since 2007 raised $41 million in funding at a value of $242 million.[4] There is a correlation between the amount of money raised and the company's value at exit – suggesting that raising more money does indeed generate more value.

In terms of companies that exited by being acquired, they on average raised $29.4 million and sold for $155.5 million. That's a pretty healthy return for entrepreneurs and investors. It takes about seven years on average to build a company to that level.

On the other hand, for the companies that exited via an IPO, it takes substantially longer – just over eight years. It also takes a lot more funding. The IPO startups raised $162 million on average before going public, generating an average of $467.9 million during the IPO process.

Of all these startups, about 60 per cent make it to the grand old age of three, with only 35 per cent making it to age 10.[5] Then there are the startups that fail. If we take failure to mean the liquidation of all assets (i.e. investors lose all the money they invested), then about 25 to 30 per cent of US startups will end up falling into this category, according to the National Venture Capital Association.[6]

Then there's the middle 40 or 50 per cent of startups that are profitable enough to keep going, but not doing well enough to become targets for acquisition. But it is from these ranks that great companies also emerge over time.

Failure is Great

We constantly hear that investors – especially American ones – love entrepreneurs who have failed. It means that they have tried, not got it right, and then dragged themselves back up again to have another go at it – this time with the benefit of hindsight.

Failure can be helpful, a real eye-opening experience – but with a big caveat. The big question is whether it was a business or personal failure. Take, for example, an entrepreneur who pulled together a solid team, raised money from good investors, built and maintained good relationships and built a great product – but perhaps their timing was off. Or perhaps they were lacking a core person in the team (and the technology failed). These kinds of failures are business failures – and can provide some harsh but useful lessons. Entrepreneurs in this category tend to attract teams and investors a second time.

Take another kind of entrepreneur. This guy is more aggressive. He takes big risks, both with the business and personally. While everything is going well, investors and employees tend to stand behind him. But if things get rocky, and some real challenges appear, then people tend to distance themselves. If that business ultimately fails, there's a much lower chance that people will work with him again.

It is worth remembering that there is a bit more to business than just financial success, and, in an interesting echo from a company fraught with a history of difficulty, how important human relationships ultimately are:

'Failure of your company is not failure in life. Failure in your relationships is.'

– Evan Williams, founder, Twitter and Blogger

Billion-Dollar Secret Sauce

But let us go back to the 0.07 per cent of companies who have achieved billion-dollar success over the last 10 years. Is there common ground that binds our billion-dollar apps and Internet companies together? Is there something those entrepreneurs know – or do – that makes them different? There definitely is.

I've pulled together research from a number of studies. I've based this on a dataset of all kinds of Internet companies[7] from all around the world, from a number of data sources, that have a value of at least $1 billion, in either the private or public market.

I chose to focus on consumer – and enterprise – companies to provide a fairer, and broader, view of the possibilities. Apps, after all, are not only focused on consumer markets but are increasingly invading big enterprises and changing the way we work.

I have also included billion-dollar tech companies as well as apps. Why is that? Because, as well as building a billion-dollar app, you're also going to be building a great technology company. App-centric companies have a number of mobile-specific challenges – and opportunities – compared with those of just Internet or technology companies. But at the end of the day, high-growth technology companies face a lot of overlapping challenges – so let's see what we can learn from our billion-dollar brothers.

IT'S A SMALL CLUB. As we saw on page 63, there are only 43 companies that are members of this rather exclusive club – and only 12 app-centric companies breathing that rarefied air. But I am going to focus on the number of consumer ones.

IT TAKES SEVEN YEARS. It's not easy to build a billion-dollar company and it doesn't happen overnight. On average it takes seven years to reach a billion-dollar valuation (and either an IPO, or merger, or acquisition), so you had better be in for the long haul. The minimum was just under the two-year mark (Instagram and YouTube) and the upper end was eleven years (Pandora). As the inside stories show, this path is never smooth. Perseverance – and a belief in the long-term vision of your company – is key if you're going to make it all the way.[8]

FIVE BUSINESS MODELS THAT WORK. It turns out that five business models underpin so many billions in value – and, interestingly, each model seems to contribute equally in terms of value. The first is gaming, where users pay for a virtual service or good. The second is e-commerce / marketplace, where users pay for a real world good or service. The third is advertising (or consumer audience building in the case where the company has not yet switched on the advertising). The fourth is Software as a Service (SaaS), whereby users pay for cloud-based software (typically via a subscription model). And the last is enterprise, whereby companies pay for larger-scale software (again, via a subscription-type model). So there isn't a huge amount of reinventing the wheel here. If you want to make it big, it's pretty clear what business models to stick to.

EXPERIENCE MATTERS. Despite all the media hype, most billion-dollar companies are not founded by youngsters in their twenties, but by people of an average age of 34 (that makes my 36-year-old face smile). Fortune also favours those who have known or worked together for many years – in fact 90 per cent of founding teams fall into this category. Thirty-five of the companies had several cofounders, the average being three. Four of the companies were founded by individuals. What does this mean for the average person? It means get out there and start your first company – that's the best way not only to get the experience you need, but also to make the contacts and find the cofounders for your next business.[9]

In addition, all but two of the companies had founders with previous experience working in technology. Also, only three companies did not have a cofounder with technology experience, which seems to make a lot of sense when founding a technology business. Eighty per cent of the billion-dollar companies had at least one cofounder who had previously started a company – often the founders had started several previous companies.[10] So, once again, you should be encouraged by this. Hardly anyone is incredibly successful first time – it's about persistence.

MOST CEOs STAY FOR THE LONG HAUL. I was quite surprised to learn that 76 per cent of founding CEOs stayed with their company all the way through to either an acquisition or IPO, with 69 per cent actually keeping their CEO role. This is a strong reflection on the calibre and experience of the founders who are building these billion-dollar companies – a formidable amount of staying power, vision and perseverance is required.[11] In the companies where the founding CEO did change role, that change, on average, was positive, generating a higher value for the company. Enterprise companies changed CEOs in 40 per cent of cases, while only 25 per cent changed in consumer companies.

EDUCATION GIVES YOU AN UNFAIR ADVANTAGE. Even though eight billion-dollar companies were founded by college dropouts, the vast majority went to top universities. Stanford leads the charge with 33 per cent of the cofounders. Eight founders went to Harvard, five went to the University of California at Berkeley and MIT had four. The majority of founding CEOs had technical degrees from university. While this may seem a little daunting, you can now get an MIT, Stanford or Harvard education essentially for free via great online course programmes such as OpenCourseWare (from MIT), edX (from MIT, Harvard and Stanford, among others) and from iTunes U.

LOCATION. It's clear that not only does the network and connections gained from going to a top educational institution help, but the network combined with the thriving geography of San Francisco makes for a killer combination: 27 of the billion-dollar companies are based in the Bay Area. New York is a distant second with three. I suspect that, over the coming years, international hubs such as London, Berlin and Paris will play an increasing part as well.

ENTERPRISE IS HOT. This is something very exciting – and ripe for opportunity. The characteristics of billion-dollar enterprise start-ups lean towards raising a lot less capital (typically, they require only 40 per cent of the investment that a consumer-focused startup does), thereby increasing returns not only for investors but for founders and employees. On the downside, it does take a little longer, on average, to see an exit event. So, for those of you with deep corporate experience, take that knowledge and those connections – the sector is ripe for innovation.

IT'S ABOUT GOING PUBLIC. Of the 10 companies that have been acquired, the average valuation was $1.3 billion. For app companies this amount was $1.67 billion (not including Whatsapp's outrageous acquisition amount as it skews the average). This represents the lower end of valuations, which makes sense, since acquirers want to get them at a lower price. So the majority of companies are exiting via an IPO. This is a great sign, because it means that entrepreneurs are focused on building companies that last – rather than just building and flipping them for a quick sale.

What Does This All Mean?

There is no easy way of building a billion-dollar app; there is no single path. Your app will have to find its own path and determine its own destiny.

But the good news is that there is a relatively systematic way to approach your path to success. Start with a big problem, a novel solution and a huge market ready to adopt it. Then build a great product that users love, and then prove they love it with data showing they are willing to pay for it. Combine that with a robust strategy that attracts users systematically and at a cost that is less than what users will potentially generate in revenue for your app.

Combine all of this with a diehard team sprinkled with people who have built companies before, and you're going to maximise the chances of building an app business that is going to last. The formula is not magic, it is simple, and it's about how religiously you adhere to the tactics and the calibre of people you attract to join you on the journey.

One of the most amazing things to appreciate – and what is a huge driving force behind billion-dollar success – is these entrepreneurs not only possess ridiculous long-term vision but also have the ability to learn on the spot, be constantly aware of their surroundings and then adapt to the market nimbly.

Before we get stuck into the deals of how to do it, let's look at the inside story of the very first member of the revered Billion-Dollar App Club. It's a window into a rather unique journey – but one that has already been repeated a number of times.

Instagram

Instagram was the first billion-dollar app – and one with a fascinating journey. I admire Kevin Systrom, Instagram's CEO, for possessing such a clear vision and such an ability to build a singularly brilliant app on a single platform. The final twist to the story – doubling Instagram's valuation in a mere four days – is the stuff of Silicon Valley legend.

Instagram marked a turning point in Internet history: it was the first billion-dollar acquisition of an app. It led to speculation that Mark Zuckerberg had lost his mind and that Silicon Valley crazy-think was back in full force. Were people deluded into thinking it was 1999 all over again? On the surface, Instagram was just a group of 16 developers hacking some software in a poky office above a pizza shop in Palo Alto. They hadn't made a cent in revenues. But they had attracted more than 30 million users in record time. And, unlike those of most of their competitors, those users were actually sticking around.

A few years after the acquisition there is hardly a soul in the business world who would disparage – or even vaguely question – Zuckerberg's decision. Instagram had well over 150 million active users[12] by the end of 2013 and the momentum hasn't stopped. It launched an advertising product for brands on 1 November 2013.[13]

So let's take a moment to dive into the details of Instagram's famous journey and understand a bit more about how Silicon Valley thinks, and why the billion-dollar acquisition of the company by Facebook actually does make sense.

The face of Instagram

Instagram has two cofounders. The first is Kevin Systrom, who is the app's best-known public face.[14] Systrom grew up in a rather nice Boston suburb called Holliston. He was a pretty smart kid, fascinated by computers and photography, which led him to enrol at Stanford – but Silicon Valley had always been on his mind. While at Stanford he snagged a summer internship at Odeo, the company that would later evolve into Twitter. It's there he made friends with Jack Dorsey, who sat at the neighbouring desk. Dorsey and Twitter would play a fascinating – and tense – role in the future of the photo app.

Systrom graduated from Stanford in 2006 with a double major in engineering and management. He then pulled a two-year stint at Google working on Gmail, Google Reader and then eventually worked for the Corporate Development team.[15] Next he worked at a travel site called Nextstop (not long after he left Nextstop it was acquired by Facebook). But Kevin wanted to do something of his own.

Whiskey and Burbn

In early 2010, Systrom launched his first tech startup on his own, and, because he was a programmer, he was able to get away with it. He saw the potential of mobile and the huge explosion of apps focused on location-based check-ins (this was when the Foursquare app was the hottest thing around), and got hacking.

He jumped on the bandwagon with a social check-in mobile website called Burbn.com. Systrom had a fondness for fine bourbons, so he followed that mantra of building something that you truly love (in name at least).

Kevin raised half a million dollars in seed funding from two venture-capital firms: the prestigious Andreessen Horowitz and Baseline Ventures. He seemed to be off to a good start. However, despite his efforts, he knew he was pushing a boulder up a mountain with his me-too site and he had a limited runway to prove himself. Its tagline was pretty uninspiring too: 'A new way to communicate and share in the real world'. I think it could apply to anything.

Cofounder inspiration

But the story of Instagram is not just of Kevin Systrom.[16] As often happens, there is another cofounder, in this case Mike Krieger. According to Kevin, despite being invisible to the public, Mike was very much the soul of the app.

Mike grew up in Brazil, and moved to the United States in 2004 to study engineering at – you guessed it – Stanford. He was the more conservative-engineer type, but possessed a strong creative and design edge. After he graduated from Stanford he joined the super-hot startup Meebo, a clever browser-based chat platform that would explode in popularity.

But what Mike really wanted was to branch out and do something new and different.

Systrom knew that he had to break out of a rut with Burbn – and turn it into something amazing. He also knew he needed a cofounder to help him to do that. Having both attended Stanford, he and Mike Krieger knew of each other but had never clicked. However, given how small the community – and coffee-shop scene – is in Palo Alto, it wasn't much of a surprise that they would frequently encounter each other.

And so, on one normal caffeine-fuelled day in 2010, he approached Krieger to join him at Burbn: 'Hey, this is going to be a real thing – are you interested in being my cofounder?' Krieger was immediately interested.

It was a bit of a risky strategy because Kevin knew he was about to do a 'bait and switch'. Kevin was planning to do something completely different once Krieger was on board: a perfect time for the greatest of Silicon Valley euphemisms – the pivot. Pivoting your business is admitting the failure of your current (and currently funded) idea, clinging on to whatever remaining investor money you have, and trying your next new idea. Funnily enough, investors (at least in the Valley) aren't too upset about this strategy. At the seed stage it's much more about investing in people, and accepting that adjustments to business models or core product strategies are inevitable (and may, as in this case, lead to a great 'new' idea).

The only remaining issue was that Systrom didn't know precisely what he wanted Burbn to become next.

Instantly professional

When you have a passionate cofounder, magic can happen. Kevin's risky strategy paid off, and Krieger was not massively surprised that he wanted to re-evaluate everything that was currently Burbn. Together they began to rip their business idea to pieces. Now was the time to be ruthless. They needed to find something new, fresh, clear, different – and something they fundamentally believed in.

So they dissected the mobile website they had – and reviewed the user proposition and features from the ground up. Given that their focus was entirely mobile, they jettisoned their mobile website strategy and opted for building a native iPhone app (their mobile website was very slow to load, and provided a pretty average user experience). Their decision was made easier when Apple launched their iPhone 4 which was markedly faster and better than the previous-generation iPhone.

But they still needed to figure out what their app was going to do. At the time, mobile and location-based check-ins were synonymous.

And then something happened. According to Systrom, something clicked: 'Instead of doing a check-in that had an optional photo, we thought, Why don't we do a photo that has an optional check-in?'

That one small idea would completely determine a new tack – and a new focus on photos. But it still didn't feel perfect. During summer 2010, when they were rushing to rebuild their service as an iOS app for the iPhone, Kevin took a short trip to Baja, Mexico, with his girlfriend, Nicole.

During a walk along the beach, Nicole told him she'd be reluctant to use the new app because her photos would never look as good as the ones taken by one of their friends. Systrom dug a little deeper – realising that Nicole thought this friend's photos were fantastic because he was using another app that would apply cool filters and effects to the photos he was taking. It was a bit of a Eureka moment as it dawned on him that filters could make a huge difference.

That night Systrom went back to their hotel room and began searching the Internet to find out how to build a photo filter. That night he coded X-Pro II, Instagram's very first filter. He and Nicole started using the filter and posted the first Instagram photo – a little Mexican dog lying next to Nicole's foot. They knew they had the beginnings of something great once the filters were in place, receiving further confidence once they had shown the app to Jack Dorsey, who was glowing in his praise.

The app was officially launched in the Apple App Store within a couple of months.

Funding

Instagram launched on 6 October 2010. On its first day, it garnered around 25,000 users. Within a few months, in May 2011, it hit 3.75 million.

By February 2011, fuelled by their strong growth, Systrom and Krieger were looking at $7 million investment from Benchmark, a prominent venture-capital firm, valuing the app at $25 million. A number of other investors contributed to the round, most notably

Jack Dorsey, angel investor Chris Sacca and Adam D'Angelo, who joined Facebook in its early days (he was good friends with Zuckerberg, who had been his roommate in high school) and went on to start a company called Quora.

Systrom and Krieger continued to execute and focus – they were laser-focused on a single platform, the iPhone, and doing a single thing, sharing photos – really well. This tunnel vision translated into very impressive 'stickiness' for the app. Not only were they able to attract new downloads and new users at a phenomenal rate, but those people were sticking around. Too many new apps would see a 'pop' or surge in usage, and then users would get bored over time and stop coming back. Instagram bucked the trend.

In early 2011, Roelof Botha, of Sequoia Capital, reached out to Kevin. 'A lot of hot startups were losing users as quickly as they get them, like people who get on a bus and then get off in the back,' said Botha. 'But they retained their users.'[17]

Botha was ready to invest $50 million in Instagram to further turbocharge its growth. It was now looking like the app was being valued at $500 million.

An attractive target

No social photo app – or site, for that matter – had had this kind of traction before and it was this incredible stickiness, and fresh1app was a breakthrough; it had understood the formula of photos, filters and social, and people couldn't get enough of it. As seasoned investors know, there are typically only one or two companies that end up dominating these types of spaces. Once consumers have decided the winner, competitors have two options: either acquire them or build their own.

Both Twitter and Facebook had been following Instagram's growth with great interest – and concern. Such growth could pose a threat if left unchecked. The Twitter connection was very close, through Jack Dorsey, Twitter's executive chairman.

There was also a historic relationship between Zuckerberg and Systrom. Not only had they met a number of times at various events in the Valley and at Stanford, but, during his undergrad days, Zuckerberg had even tried to get Systrom to drop out of college and join Facebook.

On top of this, Zuckerberg saw early potential in the photography app, having invited Systrom over to his home in Palo Alto a number of times after Instagram's launch. The groundwork for an acquisition had been covered from the early days.

Here comes the Easter Bunny

It's hard to appreciate how small the community is and how few key players there are in Silicon Valley. Relationships clearly play a large part. It was the investment interest from Roelof Botha at Sequoia that put a $500 million valuation on a company with 13 employees and no revenue, and caused interest to intensify.

In April 2011 things started to hot up for Instagram. In the previous few months its user base had doubled to 30 million, and the Android version was about to be launched (when it was launched on 3 April it added another 5 million users overnight).

Early that April, Systrom was having drinks with Jack Dorsey, and Twitter's CFO Ali Rowghani. According to Twitter, the company made a formal offer to buy Instagram for an all-stock deal that valued the app at around $500 million (which was logical) – but there was no cash component.

Dorsey and Rowghani claim that they handed Systrom an actual term sheet, something that is disputed by Systrom, who claims the conversation did take place but no formal documentation changed hands. Systrom called the then Twitter CEO Dick

Costolo on 4 April to tell him Instagram was going to accept investment from Sequoia and remain an independent for the time being.

That was precisely when things changed. Systrom communicated the same news to Facebook CEO Mark Zuckerberg – but this CEO wasn't ready to take no for an answer.

Enter Zuckerberg

The final chain of events is fascinating, and demonstrates how quickly everything moves. The final negotiations would start on Good Friday and end on Easter Sunday.

At this point, Kevin was just hours from signing a deal for a $50 million investment from Sequoia, and Facebook was only weeks away from an IPO – one that was poised to value it up to $100 billion. Given the context, it's particularly impressive that its CEO was able to respond so quickly and decisively to make the deal happen. Here's how it transpired.

On Thursday, 5 April, Zuckerberg texted Systrom, saying he wanted to talk further. He wasn't taking no for an answer.

On Friday, 6 April, Systrom went over to Zuckerberg's house in Palo Alto. Zuckerberg was adamant that Facebook was the perfect home for Instagram, and that he would do everything required to make the deal work – something that would be sealed with a huge offer. Systrom was equally adamant that it would be best for Instagram to remain independent, but still put an opening number of $2 billion on the table.

Zuckerberg then reframed the negotiation. He was planning to pay for Instagram mostly with stock, and asked Systrom what he thought Facebook could be worth. If he believed Facebook could one day match the valuation of, say, Google at $200 billion (or more), then valuing Instagram at 1 per cent of Facebook would be the right way to think about it. It was this strategy that led to the offer that Zuckerberg would eventually put on the table.

Deals happen quickly in Silicon Valley. Even though Zuckerberg owns 28 per cent of Facebook's stock, he controls 57 per cent of its voting rights, which means he can act independently, and, more importantly, very quickly. Systrom also owned about 45 per cent of his company, which gave him the sway to nail the deal without too many issues.

On the Saturday, Systrom was back at Zuckerberg's place to negotiate the final details (with their lawyers). Apparently the process was interrupted by a party Zuckerberg was throwing for *Game of Thrones* – apparently his favourite show.

On Sunday morning, 8 April, Zuckerberg, alerted his board of directors that he intended to buy Instagram.[18]

The final number – agreed on the Sunday – was $1 billion. It was a combination of Facebook stock and a dangled carrot of $300 million in cash. The staggering offer was double what Sequoia and Twitter were valuing the company at. Though the numbers the two talked about were eye-watering to any normal human being, the real deal clincher was centred on the fact that Instagram would be able to operate independently within Facebook – something critically important to Systrom and his team.

> 'I'm not sure what changed my mind, but he [Zuckerberg] presented an entire plan of action, and it went from a $500 million valuation from Sequoia to a $1 billion [one from Facebook]'
>
> *– Kevin Systrom*

Clearly, both the sheer amount of money and the natural fit with Facebook swayed Systrom.

Aftermath

On Monday, 9 April, Instagram's deal with Facebook was made public. Both Jack Dorsey and Twitter's CEO Dick Costolo were left wondering why they hadn't even been given the opportunity to put in a counter-offer. Systrom closed the funding deal with Sequoia before Instagram was acquired by Facebook. That gave the venture-capital firm an instant return on its investment.

'I have to give Kevin a lot of credit for keeping his word,' said Botha.[19] Apparently the two had had only a handshake agreement – one that cost Systrom a personal fortune to honour.

When Facebook did finally IPO, the final value of the deal was $736.5 million because of the initial drop in Facebook's stock price, but, in the period since the IPO, Facebook's stock price has bounced back to the IPO level and proceeded to rise even further, so Kevin Systrom did indeed make a fantastically good deal.

Andreessen Horowitz, the firm who invested the first $250,000 in Burbn, made a return of $78 million when Facebook bought the app. That's a rather healthy 31,000 per cent return. Nice.

STEP 1

The Million-Dollar App

Building a Founding Team, Validating Your Product and Raising Seed Funding

Going from Nothing to a Million-Dollar App

Idea

- You have a big idea and you've validated that millions, if not billions, of people will love it. Get out there and test your great idea for an app, put it on paper, and start getting feedback. That is the first step.

App

- Figure out how your idea – your solution – translates into an app. Sketch on napkins, sketch ideas in Photoshop, do whatever you need to do to make your idea real and communicable to others.

- As you flush out a great design, start prototyping it. Your goal is to get it into user's hands quickly, so that you can get as much feedback as possible. Your goal is to get to drive towards delivering wow!

- No matter what happens, you'll need to have an app ready for real users' hands. If your business model is simple (gaming, Software as a Service) expect to be operational, at least in a basic way.

- If your model is more complicated (marketplace), then you have a solid proof of concept. Use that to secure investment to build it out.

Team

- You should not use the excuse of not having a cofounder to slow your progress, but finding a partner in crime who shares the vision and has complementary skills will make the journey more enjoyable.

- Earmark key people you need in your team to make it work – and do everything you can to get them on board. Recruiting will be your number-one job throughout the life of your app.

- Three key roles need owners: someone must lead the product vision; someone needs to build the technology; and someone needs to be focused on getting users and generating money.

Users

- Your mum promised to use your app – so make sure she does. Persuade every friend and family member to download it, use it – and give brutal, honest feedback.

- Figure out your target user group early, and find out how to target them and get the app in their hands.

- Aim to get feedback from hundreds of real users, if not a thousand (if you're a marketplace model a couple of hundred will be good going).

Business model

- Gone are the days of not having a business model on Day One. There are only five business models that power all billion-dollar apps – make sure you know which one will power yours.

- Think about analytics and how you're going to measure the performance of your app, and therefore the performance of your business.

Valuation

- Ideas are a penny a dozen. It's execution that's worth money. You'll drive a solid valuation if you get the right team together to deliver a product that users love to use. Do that and you'll be worth $1 million. If you do that very well, you'll be worth around $4–5 million[1] – the average valuation of a funded startup at this early stage.

Investment

- You're investing your own blood, sweat and tears at this point. If you nail your idea, team, prototype and some users, then someone will probably be happy to give you anywhere between $250,000 and $1 million in exchange for owning a percentage of your new app company.

Chapter 5

Let's Get Started

The start of any journey to developing a hugely successful app business is a billion-dollar idea and a dream of making it big. Through this first part of the journey it's all about validating your idea, the market, your product, the basics of your business model – and putting together a plan of attack. You're putting together an end-to-end plan for your business, and supporting it with research, data and innovative thinking. At the end of this section you'll have real customer validation and metrics, and be confident that your basic app will become a business. The ultimate validation comes when experienced people buy into your vision and invest money – whether those people are friends, family, cofounders or investors. If you can convince the last group especially, you're probably on to a good thing.

Once you've combined your huge idea with a huge market, it's time to find a cofounder. It's worth considering now because the earliest days of your app are about challenging, distilling and refining your idea into a service, into an app. For just about all of us that is better done in the context of a team – a team most powerful when complementary skills are brought to the table.

The next step is creating the identity for your business – its name, brand, style and first designs. This is the genesis of your app: what it's going to be called, how it feels and how it's going to engage people in a very strategic way.

Once these big questions are answered – and decided upon – it's a lot easier to move to the prototyping phase. This is when the app itself takes shape: you start thinking about details such as precisely what every screen looks like, what every button does and what service or experience you want to deliver to your users.

Finally, it's about validating your app with users – in the real world – and starting to gather objective feedback. It's about measuring – with numbers – how much people love your idea, how much people love your prototype, how much time they spend using it – or how much money they would spend on it.

If you've been able to bootstrap your way to this point – existing on the smallest amount of money possible probably while holding down your day job or living in your parents' basement – then you'll be close to justifying a million-dollar valuation for your app. That's real objective validation. That is the basis on which people will invest in you – and your idea.

That is the beginning of your company.

How Badly Do You Want It?

This book is not a simple 'how to' guide about building a billion-dollar business. It would be naïve to think that you could encapsulate everything required to deliver such success in a single book. Rather, there is a remarkably common set of tactics and strategies that the top entrepreneurs – especially the new wave tackling the mobile challenge – all share. Going through this journey has opened my eyes about how complex, lucky and tenacious you have to be to make it to the very top.

The most important question you can ask yourself, as an

entrepreneur, is, How much do I want to sacrifice to achieve success? Everyone wants to be successful, everyone wants a multimillion-dollar exit, everyone wants to be well respected for what they've achieved. Not everyone, however, is willing to work 80-hour weeks, or be told by everyone they meet that their idea won't work and still persevere every day, or constantly sell their idea to every person they meet to build up their team, or work through mountains of legal documents late into the night, or spend months chasing investment to keep their idea alive.

Understand from the onset that it's a hard – and long – fight. As we saw earlier, if you're going to reach a billion-dollar exit for any tech startup it's going to take you on average seven years.[1] So buckle up. If you're willing to sacrifice and persevere, then it will be the most exhilarating journey of your life.

It's Not Just an App: It's a Business

One of the first questions you should be asking yourself is, How will my app make money? Understanding potential business models is your key to real success. In our family of billion-dollar-app winners – such as WhatsApp, Uber and Square, and games such as Candy Crush Saga, Angry Birds and Clash of Clans – the companies that generated revenues from Day One are the ones who wield the real power, create the most value and, most importantly, control their own destinies.

Unfortunately, we no longer live in the wonderland of the year 2000, when it was only about users and traffic – and revenues didn't really matter. Ah, those were the days! In 2007, when we raised our first round of financing for my own startup, WooMe – a video dating website – we were able to find investors with a speculative business model, which seemed like a good thing at the time. But when we eventually had to transition from an entirely free dating site to a 'freemium' one – where basic features were free, and users paid real cash money for features such as video chat – it was a difficult process. And, during the struggle to deliver the business model, we were very much beholden to our investors. Trust me when I say it's no fun being in that position.

'But what about Instagram and Snapchat?' I hear you cry! While it is indeed *possible* to build an app with bucketloads of users and no revenue to speak of, it's not a strategy that you should bet on. What if your app doesn't take off? No revenue means fewer options, which means less negotiating power, and ultimately a lot of unnecessary risk.

Don't worry, though: your business model doesn't have to be perfect from Day One. But my advice is to make sure you have one – build a business, not just an app. Throughout your evolution from a million-dollar to a 10-million-dollar to a 100-million-dollar app and beyond, it's very likely your business model will evolve. I'll lead you through some excellent examples where not only have some apps been able to adapt successfully, but they have also become category leaders because of their ability to adapt *faster* than the competition.

Today the expectation is very much that your app should generate revenue from the start. Why is that? Mainly, because it is perfectly simple to do so. In addition to services provided by Apple's App Store and Google Play to allow your users to pay with a single tap from their iTunes or Google accounts, there are numerous other ways to make money. We'll go through the most powerful ones right now.

Five Core Business Models

So there's a great way to zero in on the top business models: focus on the existing billion-dollar apps and Internet companies, and then explore the models they have used to get there. As explained on page 63, since 2004 there have been over 40 companies that have been valued at $1 billion or more.[2,3] It's an approximate number because private companies can be tricky to put a valuation on, but trust me for the moment, and I'll explain how I arrived at those numbers towards the end of the chapter.

I am going to concentrate on five business models that scale nicely into the billions. The companies referenced below – all of which are mobile-first, or have a massive app channel – are valued above $1 billion, and some a lot more.

1. GAMING. The Apple App Store and Google Play have been responsible for the growth and revenue-generating capabilities of a lot of apps. From pay-per-download to in-app payment and subscriptions, I am counting these as a grouping of business models that work great, especially for gaming apps, and anything virtual that can absorb the 30 per cent commission charged by Apple and Google. Angry Birds, Clash of Clans and Candy Crush Saga are all in this category.

2. E-COMMERCE/MARKETPLACE. I am lumping e-commerce and marketplace together for simplicity. While naturally there can be huge differences between the two – network and other effects – in terms of revenue generation charging payments for transactions for goods or services via credit card (or other payment methods such as PayPal) is relatively standard. Mobile players such as Uber, Hailo and Square fall into this category as well as more traditional players such as Amazon and eBay.

3. CONSUMER AUDIENCE/ADVERTISING. Audience building is a valid – though very tough – strategy. Players using this model are all about getting big, fast – and then being as sticky as possible. Our friends Instagram and Snapchat fall into this group. Flipboard represents the flipside that actively targeted big-ticket advertising from the get-go –

and has been able to deliver. These players either develop their own advertising platforms (Flipboard), or seek to be acquired and monetise off the acquirers' already well-established platform (as Instagram did with Facebook).

4. SOFTWARE AS A SERVICE (SaaS). This is where consumers pay for cloud-based software or services. WhatsApp is the first 'mobile-first' company with this model, charging some users $0.99 per year as a subscription via app-store payments, though its business model has evolved a number of times. A number of companies – which are not mobile first, but still have heavily used apps, such as Evernote (a note-taking app), Dropbox and Box (two document storage apps) – have this business model, whereby they charge subscriptions either via app-store payment channels, or bill you directly via your credit card.

5. ENTERPRISE. This is where bigger companies pay for larger software solutions. In many senses this is an extension of SaaS. We haven't seen any app-first companies adopt this business model, but there have been 10 billion-dollar Internet companies since 2004 that fall into this category, including Workday (an on-demand human-resources management tool), FireEye (a security company), and big data companies Splunk, Palantir and Tableau Software.

On many levels I think this is pretty reassuring. It means that coming up with a robust business model is not rocket science. All five of the above business models have yielded fast-growing Internet or app companies; they have delivered strong and robust profits; and they are all pretty simple to understand.

The even better news is that each one of these business models has become easier to actually put in place (in terms of the payment or monetisation details) – and there are more people who understand how to apply them in practice.

Attracting the Best

Standards are continually getting higher. In order to build a truly great app and business, you're going to need to perform well on all fronts. In order to attract the best people across all parts of your organisation, and even in the case of wooing a cofounder, it helps to have a robust business model from the onset. Ensuring that your app is inextricably linked to one of the above five models is a tough exercise but ultimately one that pays dividends (obviously).

At the same time, your ability to attract great investors – especially early on – is only improved by having a business model in place from the outset. While some businesses have demonstrated the ability to avoid such a necessity, they are few and far between. And remember, attracting investment is definitely not a final goal: it is merely a way to expedite your ambitions. Your real goal is to build a sustainable and ambitious business.

As we go through the early steps of getting your app off the ground it's worth thinking about the snowball effect: all the decisions you make early will form the basis and core of future decisions. If you adopt the mentality of disciplining yourself to work on developing the best business model, conducting the best research, hiring the best employees and seducing the best investors, then you're laying the best possible foundations for future success.

The Cofounder Dance

Starting and running a company can be a lonely business. It is hard building a vision that others don't (yet) share; it is hard building a business with little money. All companies go through ups and downs, and just as with anything else in life – good or tough – it's more pleasant to share it with someone.

The many challenges of growing a startup are better faced with a cofounder: long and lonely nights of working behind your laptop are a lot more enjoyable with someone else; it's easier to stomach the depressing news that another investor loves your idea but the opportunity just isn't right; and, equally, it's more exhilarating to share the success of your millionth download with someone who has worked equally long and equally hard to achieve it.

As we saw earlier, 35 of the tech companies that have achieved billion-dollar valuations since 2004 have had two or more founders – with the average number being three.[4]

While it's preferable to have a cofounder, you don't necessarily need one, and you shouldn't use not having a cofounder as an excuse not to start the business you always wanted. That's a copout.

Three things that make a great cofounder

Finding a cofounder is a lot like dating. And, if you're serious about making your startup a success, you should realise that this is going to be a 4–8-year commitment on average. That's about as long as most marriages, the average duration being about eight years.[5] So this is not something to take lightly.

So what should you be looking for in a cofounder?

CHEMISTRY. Above all, you need to get along. In an ideal world your cofounder will be someone you have worked with before. You know how they operate, your personalities gel, you can bounce ideas off each other, and you find each other energising. You need to like spending time together – since you're going to spend a *lot* of damn time together. And, above all, you need to trust and have faith in each other. You're not going to succeed if you feel as if you have to micromanage each other, or if you feel as if you're both motivated by different things. Out of the billion-dollar success stories between 2004 and 2014, 90 per cent of the cofounders had at least a few years working together, or knew each other from school.[6]

COMPLEMENTARY SKILLS. Opposites attract, and in the startup world this couldn't be truer. When you're starting a company, do yourself a favour and make sure your cofounders don't have overlapping skills – you don't need three master's degrees in business administration (or MBAs) to start a company (in fact, that's probably a recipe for disaster and PowerPoint suicide). In the early days you need someone to take care of the product and business side, and someone to build the software. If you're great on the business side, you'll need to stretch yourself to develop a product that users really want to use, and then seduce a great software developer to work alongside you. If you're a great developer, make sure you hunt for someone who is dedicated to producing a great business model and a great product.

The key team skills that you're looking for are the ability to discover, design and refine a great product concept (and business model) and the ability to translate that vision into actual software. With those bases covered, you've cracked one of the toughest parts of getting started.

PASSION. Being an entrepreneur is a vocation. If you've ever been stuck in a job thinking you could do your boss's job better, or have said to yourself, 'What the hell is the CEO thinking? I'd never do that in his position', then you should test yourself and start a business. It doesn't matter whether your cofounder comes from a business or marketing background, or an engineering or software background, but it *does* matter that you both have the same level of passion. Make sure your cofounder has the same drive to solve a big problem, to change the status quo. Without the passion, without the drive, you're never going to overcome the myriad challenges that will appear on the way.

Some founders are more equal than others

Before you start looking for a cofounder, you need to think carefully about how much control of your business you are willing to give up in order to bring a partner on board.

At some points your startup may feel like George Orwell's *Animal Farm*. I have definitely experienced the situation where some cofounders bring more to the table than others. If you're on the more experienced side of the equation, then remember that you don't always have to go 50–50 into a business. If you're bringing a well-developed idea, money, or the basis of an app that already has traction to the table, take advantage of your position.

The very title 'cofounder' is a powerful asset – and can help you negotiate with potential candidates. Retaining equity where and when you can will pay off down the line. This is a lesson that I have learned from a number of serial entrepreneurs. They know that, as you grow a company, you will invariably need to raise money to help you grow. And, in exchange for money and investment, you will have to give away equity and control. So give away only as much equity as you need to – and no more.

Hailo early days

I joined the Hailo team when there were only a handful of people on board. There was no app yet, only a name, logo and enormous vision. The vision was to enable anyone to hail a taxi in any city in the world via their smartphone. It was a simple – and universal – proposition, but, as you might expect, monstrously complex to deliver.

Hailo was the brainchild of Jay Bregman. He was 32 when he founded the company. In Part I, I talked about Jay's first company, eCourier. Jay is an energetic visionary – his brain runs a million miles a minute. He knew that he needed someone a bit more measured, methodical and experienced by his side. So it was no surprise that he lured Caspar Woolley, the former COO of eCourier, to his new venture. Caspar brought the grey hair, bringing years of experience managing operationally heavy organisations such as Avis, the car rental company.

Next, Jay wanted someone with some deeper financial experience and clout. He was introduced to Ron Zeghibe, a seasoned executive who had been chief executive of Maiden – an outdoor-advertising company – and spent much of his earlier career in the private-equity world. Ron had had a very successful career and was by no means in search of something to complicate his happy life. Jay, however, can be very convincing, and within a few weeks a founding team of three was in place.

Six is better than three

Jay's vision was big – from the onset he wanted to revolutionise an entire industry. On the surface it was obvious that taxis were used in more or less the same way around the world. In order to capitalise on this enormous market – quickly – Jay realised he needed to get deep into the minds of taxi drivers.

This laid the way for the other three Hailo cofounders (yes, there are six cofounders in total) – three London cabbies. Not only did it make sense to have taxi drivers as part of the core team to ensure the product was focused on delivering a great experience for drivers, it would also communicate an air of credibility: that this app and this company really understand their users. It's worth noting though, that six cofounders is an exception; two or three is more the norm.

Once this founding team had been put in place, it was a relatively simple process to get out there and talk to countless drivers and validate the idea. It was a huge advantage from the very beginning. And later it would prove to also be a fantastic public-relations tool.

Wherefore Art Thou, Cofounder?

Unfortunately there isn't a magical place to find your cofounder. The good news, however, is that today there are many more events, meetups and even websites that help facilitate this process. First, remember that the best place to find a cofounder is the company you're working for right now. Scour the people in your company, pick out the people with vision and technical talent and who are as frustrated as you are with the status quo (but make sure you don't violate any non-solicitation clauses you may have signed).

I found my first cofounder when I was working at British Telecom. He was my boss at the time. We were both keen to do something bigger and better. The best bit about finding a cofounder at work is that you have a history of working together, you spend a lot of time together, and therefore you can get things rolling very quickly.

That's precisely what Stephen and I did. We pulled longer hours for a number of weeks, flushing out product designs, technology prototypes and business plans. Finally, we were convinced we were on to something huge. We recruited a great engineer from another part of the company, and then promptly quit to work on our startup in earnest, relying on just our personal savings.

It's a tried strategy that works. The cofounders of WhatsApp worked together at Yahoo! for many years before coming up with their own idea, and then promptly quit to start their new company.

The cofounders (all six of them) of Supercell – the company behind Clash of Clans and Hay Day – all worked together at another games studio before leaving to found their own. It was a similar story with the cofounders of Angry Birds's maker Rovio, and the same for the Israeli cofounders over at the social-traffic app Waze.

If that strategy doesn't work for you, or isn't an option, then get out there into your local startup community and attend any events you can. The startup world is very social, and most events involve pizza and beer and everyone is interested in meeting everyone else.

Here are some great events that will get you talking to the right people.

DEVELOPER MEETUPS. Tech companies host all kinds of events to showcase new technologies or features to developers. Facebook has one called Developer Garage. Each month the company invites developers together to talk about the latest features, such as Social Graph, Facebook Search or Facebook advertising. Companies such as Google, Yahoo!, LinkedIn and numerous other big software companies organise similar events all the time. Even though they are targeted at software developers, don't be scared to attend if you're not technical. Go along, pretend to be a developer (to get in), enjoy the pizza and beer, and then start talking to anyone and everyone. Not only will you get a flavour for what's hot in terms of technology, but you'll also start to understand how software developers think and communicate. You may even be inspired by the new features and apps that other people are building.

TECHNOLOGY CONFERENCES. No matter which city you live in, you'll find a handful of great technology conferences. From TechCrunch Disrupt in New York, San Francisco and Berlin, to LeWeb in Paris, to DLD in Berlin and Web Summit in Dublin, there are countless conferences in most major cities. Go there, mingle, talk to people. And make sure that you have 'Founder' in big letters on your name badge (people are more likely to talk to you if they think you're important or have your own company). If you don't have a company yet, don't worry. Make sure that you're prepared to talk about an industry or technology you find interesting – and then engage with people about that subject. If you do have an app idea already, then make it sound like a company (one that's about to launch), and make sure you've got a snappy two-minute explanation up your sleeve about what it does and why it's going to be huge.

STARTUP WEEKEND. This is a great event, and is designed for those of us who are time-poor. Think of it as an end-to-end business plan and software prototyping competition, crammed into a weekend. It is now operating internationally, with thousands of successful weekends completed. The concept is simple: turn up on Friday night with or without a great startup idea; people who have ideas pitch them; everyone else listens. By the end of the night everyone assembles into more or less evenly sized teams, with more or less balanced skills (the organisers do a great job of making sure the right composition of people is selected for each weekend). Over the course of the weekend, the teams refine their ideas and pitches, put together a presentation and then hack together a basic proof-of-concept piece of software by Sunday afternoon. The final team presentations happen on Sunday evening – and a winner is crowned. It's a great way to spend a weekend and meet some amazing people.

MEETUP.COM. As well as being a great startup itself (Dom Preuss, a good friend of mine, used to head up the product development team there), Meetup.com is a platform that allows anyone to create and host any type of event. There are meetups for novice entrepreneurs, for experienced entrepreneurs, for software and app developers, and everything in between. It's very much a bazaar – full of mixed-quality events and people. It might take a bit of trial and error to find something great, but the point is that it has so many events that there will be one almost every week for you to attend, which means you have no excuse for not getting out there and meeting people!

TECHCRUNCH DISRUPT HACKATHON. A hackathon is where developers (and designers and people with ideas) all get together and build something functional – a website, an app – in a fixed period of time. In this case it starts on Friday night and culminates with public demos on Sunday afternoon. With more than 800 people taking part – twice a year – in both New York City and San Francisco, this is also a great event.[7] You get great exposure to talented – and motivated – people and, if you demo something great, you get attention from not only others in the crowd (to build up your team) but investors who are waiting in the wings to invest in your nascent venture.

ANGELLIST. One of the most promising places to find cofounders – as well as anything startup-related – is AngelList. This site was born out of an entrepreneur's frustration with the complexities of raising angel funding in Silicon Valley, and is now the largest site of its kind in the world. Apart from being a great source of accredited angel investors (which I will talk about later), it has a rich jobs section, which lists a wide variety of jobs with startups all around the world, and also features people searching for cofounders. Remember that first impressions count, so make your profile as impressive as possible.

If you'd like even more suggestions about great cofounder events, meetups and websites – just visit the Billion-Dollar App website at mybilliondollarapp.com.

The right one

So you've met someone you think might be a good cofounder. Unless you've worked with them on a business level before, you need to figure out whether they are going to be the right partner.

I speak from experience here. I decided to cofound a company with a friend, who had also worked in a couple of tech startups. Unfortunately, despite knowing each other socially, we hadn't worked together and there was a clear disconnect over level of effort we each should put in and differences in technical ability and experience. In short, the venture ended up costing me a couple of thousand pounds in legal fees to resolve. So be careful – vet people upfront and do reference checks – even if you already know them.

Unless you already have an existing relationship with your cofounder, I strongly suggest that you both work on an extended project together to ensure that this person is someone you *should* be committing to for the lifespan of your app.

Here are a few things you absolutely should do before taking the plunge with this person.

GO OUT TO DINNER. Make sure you can hang out and talk about what both your visions are for the future of the company. Do you both want to build a billion-dollar app? Are you both interested in building a company with hundreds of people? Are you both willing to work nights and weekends for a year to realise your dream? Or is this person a fair-weather entrepreneur? Talk about the hardest things you've worked on before. Talk about perseverance and stamina. If at this stage your visions are not 100 per cent aligned, there is very little chance it will work.

A great example of this approach was when Mark Zuckerberg was 'interview-dating' Sheryl Sandberg for the number-two position at Facebook. While other people were involved in the founding of Facebook, Zuckerberg was without any doubt the one driving the bus for the first few years. It was clear he needed the equivalent of a cofounder – a leader and manager (and adult) – to help him take the company to the next level. He knew that he needed the perfect person to take the reins as the chief operating officer and transform Facebook from a startup to a well-oiled Internet goliath.

Since there was so much at stake with this role, he needed to make sure he was making the right decision. Over a number of weeks, he would meet Sandberg on a regular basis – at his own house, where he would cook dinner, and also at Sheryl's place, where they would philosophise about the vision of where Facebook should go. While Sheryl had a close-to-ideal CV, Zuckerberg needed to know that she shared his vision, that their communication style would work and that they clicked. It was not a rushed process, and, most importantly, it would be 'right' only if they both wanted to do it.

CREATE A PROOF OF CONCEPT TOGETHER. Another great way to verify that you have cofounder chemistry is to work on something very concrete together. So take your cofounder for a test drive. Sit down over the course of a few weekends and work on a proof of concept for your app. Start flushing out all the features of your app: the business model, the target users, how you will home in on your first thousand. Work through all the details together and see how it goes. Were you productive? Did you argue? How did you resolve the issues you didn't agree on? Did you both maintain interest and dedication to the task? Or did you both just lose interest? Setting a clear and simple goal and trying to deliver it together shows how committed you both are.

This is precisely the process that I went through with Stephen when I cofounded WooMe. We set ourselves the task of putting together both a business-plan outline that we would take to a couple of investor friends, as well as a detailed product design. Week in, week out, we met and worked through our plan, split tasks and then delivered what we both promised – and never missed a deadline. When we worked together we set goals and deliverables. We certainly didn't agree all the time – but we did decide that arguments should be solved by data and research. That was a constructive and objective way to deal with differing viewpoints. When it came to fundraising, the process was straightforward because we were in agreement and trusted each other. That initial proof of concept turned into a 10-million-user dating website with teams in London and Los Angeles and over $17 million in investment.

Red flags

It's also worth mentioning a few cofounder red flags. You'll most likely come across both people who sound compelling, but who are not, and people you just don't want to surround yourself with.

One of the most amusing types is the 'expert who wants to charge you fees upfront'. At first these 'entrepreneurs' or 'advisers' can appear quite compelling: they seem to have knowledge of the sector and claim close relationships with impressive people and a general sense of knowing what's hot. They get very excited about your idea and start suggesting a number of ways they can help you. At this point they start talking about people they have either helped to secure funds or introduced to investors, or exits they have facilitated. They start giving you the sales pitch. And then they mention that it would be worth booking a session, or an appointment, and figuring out how you can formalise a partnership or working agreement. At this point alarm bells should sound. Here's the rub: the real movers and shakers in the Internet world are more than happy to help you for free. People in the know will always help you – but typically they will do that only if they think you're a person with the skill and motivation to make your company work, or if they genuinely think that you have a solid, unique idea that is worthy of further development.

Another type of interesting person I have come across a number of times now, and am seeing more often, is the professional business adviser or coach. This person seems to be quite credible – they often have a website, a big network of other business coaches and advisers – and usually have a couple of deals to their name. Whatever you do, don't get too involved with these types. The only real advisers and coaches are other entrepreneurs.

I had an interesting time delving a little deeper into the operations behind some of these less-than-professional operators. I know a number of first-time entrepreneurs who have been victims of their typical approach. These 'coaches' start by offering their business coaching skills, then they throw in their ability to help with legal and investment advice. Instead of asking for direct payment, they ask for payment by taking equity in your business. Needless to say, no reputable lawyer or adviser works in this way. Lawyers often participate as seed investors – as do many other professionals – in a standard, transparent and above-board way. If you want someone as an adviser, invite them to join your board of directors and give them equity. The going rate for a high-profile person on our board is about 1 per cent. So beware of an adviser asking for 10 or 20 or 30 per cent of your business and wanting a passive role.

If you find yourself continually passed over or not introduced to other people or investors, then start with a bit of introspection and get honest, frank feedback from someone you trust. Once you address the glaring issues, people will be a lot more motivated to make introductions. Don't fall into the trap of thinking that paying a second-rate snake-oil salesman will solve any core issues your business may have.

Successful matches

Some of the most successful cofounder combinations involve bringing together great business, product and software experience from the very beginning. This creates a powerful, stable and self-reinforcing founding team. That means the critical aspects of the business are given the rigorous attention that only a cofounder can truly bring.

Some of our billion-dollar apps have particularly well-balanced founding teams. Here are a few examples.

SQUARE. This payments app is mainly associated with Jack Dorsey – but it has two lesser-known cofounders: Tristan O'Tierney and Jim McKelvey. McKelvey – who is a glassblower – recounted the story about the loss of a big sale because he had no means of accepting a credit-card transaction to his friend, Jack Dorsey. In February 2009, Dorsey, McKelvey, McKelvey's wife, Anna, and a friend named Greg Kidd drove north of San Francisco to a restaurant called the Pelican Inn. They spent the evening debating whether they should start a company based on the idea that the world needed an easier way to make payments in person. McKlevery became the brains behind the hardware of Square, designing the glossy white card reader. Tristan O'Tierney was the company's first iOS engineer, building the first Square iPhone and iPad apps.[8] So the team brought together business, software and hardware from Day One.

SNAPCHAT. You meet great friends in university, and that's where Evan Spiegel met Bobby Murphy and Reggie Brown. All three were Kappa Sigma fraternity brothers at Stanford University. While still at university they worked together on an app called Picaboo. Later, Evan and Bobby incorporated a new company, and created a new app under the name Snapchat. Reggie was not part of the new venture. The story sounds rather reminiscent of the Winklevoss twins, who launched ConnectU[9] – a social networking site for Harvard students – with technical help from Mark Zuckerberg. Not long afterwards Zuckerberg would launch Facebook without the twins' involvement. In any case, Bobby Murphy brought a strong engineering background to the table, and now serves as the CTO of Snapchat. Evan, on the other hand, possessed more product-development and business skills and is the CEO.

FLIPBOARD. The founders of Flipboard, Evan Doll and Mike McCue, met on what was essentially a blind date at a coffee shop set up by shared friends. They instantly knew they wanted to build something together, but they weren't exactly sure what. The company grew out of a thought experiment: what would the Web look like if it were redesigned today and were to focus on presenting content in the best possible way?[10] Mike and Evan knew that the product would need to be social – and beautifully designed. Evan was naturally suited to leading the software development (he was a senior iPhone engineer at Apple) and Mike headed up the business side of things (he is a serial entrepreneur – his biggest exit being TellMe Networks, which was acquired by Microsoft for $800 million).

The real stories behind our billion-dollar-app cofounders are pretty simple – and often pretty predictable. There is a formula – basically conspiring with people you have either studied or worked with. It's a combination of a significant amount of time working together and plotting to do something new that will sow the seeds of your company. If that's not possible, it's up to you to put yourself in situations where you're going to expose yourself to the complementary people you need in place to start your company.

Chapter 6

Solving the Identity Crisis

We live in a world where our attention is torn in numerous directions simultaneously, and where hundreds of thousands of apps are clamouring for us to hit that download button. Existing, competing and winning in this kind of environment requires great execution on numerous levels – and that starts with having a crisp name, logo and proposition. All these assets will help to elevate you above the noise.

Creating a robust name and brand from the very beginning is critical. But it's damn hard. And it can take quite a bit of time to find something that feels just right. From the onset I would recommend a couple of strategies. First, don't let the name hold you up from designing your app or even beginning the software development – you can refine the name in parallel.

Second, don't settle for an OK name. A great name is 10 times better than a good name. So invest the time now, because down the line there will be a million reasons why it's close to impossible to change it.

First Impressions

When we meet a new person, enter a new situation or even read about a new app, we are subject to the influence of first impressions. Our brains start making decisions instantly; we make judgements, many of which are based solely on our subconscious. That means someone's first experience of your app – the name, the logo, the icon, the tagline, or even the domain name – i.e. its website – can all have an influence on what people *think* about your app, before they've even tried it out.

At the core of this is your app's name. It is central to the conversations people have when they talk about how good – or bad – your app is.

So let's list the important factors that you need to consider when coming up with a name:

- Is your name short, catchy and memorable? Hailo is about *hailing* a taxi. Snapchat is about chatting, *rapidly*. Waze is about finding the best *ways* through traffic.
- Is your name distinctive? Uber meant little until it became your on-demand chauffeur. If you can't cut through the noise with a clear name, make one up. Etsy was a nonsense word, but is now synonymous with 'marketplace for handmade goods'.
- Is your name clever? Does it make people smile? The Square app allows you to 'square up' your bill. Pinterest allows me to 'pin' all the things I find interesting.
- Can your name become a verb? *Hailo* me a taxi, *Google* that word. This is one of the most powerful characteristics – and one you can't force. But you can ensure that your name is conducive to this usage.

If you manage to get all these characteristics in place, you're off to an amazing start. It may not be possible to integrate all these things in your name, but do remember that the time you invest at these early stages pays off down the line. A great name and brand are huge enablers for your app to become truly iconic.

Finding a great name

Frankly, I have found that drunken conversations involving a group of friends brainstorming (with shots as prizes for good ideas) tend to be the best – and most fun – generator of fun names and more original ideas. But, if that's not your thing, you can check out a variety of online services that will help you through the process.

One thing to remember is that the final name of your app is quite closely coupled with your domain name. Why? Because one of the main interactions between you and your users is your website – people will Google it, or click through to your website from an article or review. So the process of name generation will invariably involve checking whether a version of your app name is either available or can be purchased for a reasonable amount.

There's a great variety of tools and services online that can help you solve your current identity crisis. Not only do they help you come up with great ideas, but they also check to see whether there is a decent corresponding domain available.

NAMESTATION.COM. NameStation is a powerful combination of word-generation tools – from random-word generators (with character, special symbol and number options) to suffix and prefix tools – that all go and search the availability of the dotcom and other top-level domains in real time. Very handy. Imagine you're building a healthy-food app. You might enter words such as 'tasty', 'healthy' and it helps by generating more, similar, words, and then immediately checks whether domains such as tastyapp.com, healthyfood.com and hundreds of other permutations are available.

SEDO.COM. This is probably the largest – and most usable – *premium* domain reseller on the Internet. In one central place you can see all the great domain names you'd kill to own that enterprising individuals have already purchased and are now trying to resell (to entrepreneurs like you) at an inflated price. It's vexing – but buying domain names is a just land grab. The site is easy to use – and reputable. I've purchased domains here before and had a flawless experience – but, as mentioned before, be ready to part with a few thousand dollars to get something short, catchy and memorable.

DOMAINNAMESOUP.COM. This is a pretty fun site. For those looking to grab a short and unique name such as etsy.com, it helpfully lists all the available three-, four- and five-character domains – and also the ones that are currently for sale. It also has a big selection of tools to generate all kinds of company and app names for all kinds of industries and segments. I've used it many times and recommended it – with great feedback – to a lot of people.

INSTANTDOMAINSEARCH.COM. This is a simple and handy site that allows you quickly to type in domain-name options and check whether they are available in real time – no page refreshes. It's very handy. It also gives you a small selection of similar and related domains.

Domain names

For a technology company, a domain name has impact. Think of it like owning great high street real estate. If it's simple, clean and easy to remember it will lend an air of credibility to your business and build trust with your users ('Wow, if they own that domain they have been around for a while!'). That means that getting your hands on the dotcom version of your domain – or something pretty close – can be very valuable.

Square, for example, wasn't able to initially acquire square.com – it was priced in the millions of dollars. So, instead, it settled on squareup.com. Today – even after purchasing square.com – the principal website is squareup.com. The concept of 'squaring up' is a more powerful brand message that just the four-sided shape.

In terms of URLs, here are some golden rules.

- Owning the dotcom version of your name is ideal. It confers credibility, trust and a clear leadership position. Launching with a 'national' domain (such as .co.uk or .fr) is fine if you have small ambitions but won't cut it if you're serious about billion-dollar status.

- Get as close as you can to the pure version of your domain. Starting with variants is perfectly fine: Twitter operated on twittr.com for many years before having the money to purchase twitter.com; similarly Facebook was thefacebook.com for many years; and Uber had ubercab.com before it acquired the sleek uber.com domain.

- Starting out with a domain name like '[companyname]app.com' is great strategy if you're focused just on the mobile side of things.

- Other alternatives such as 'get[companyname].com' and 'get[companyname]app.com' are also good ideas to get the ball rolling (ideally, you'll be able to purchase a cleaner version down the line).

- And, if you're serious, start the process of communicating with the person who owns your 'dream domain' – it may take months, or even years, to eventually acquire it.

It used to be true that you could find a great domain name for $10,000, but that's frankly pretty hard. Fred Wilson, a very seasoned investor behind companies such as Twitter, Tumblr and Etsy (he also invested in Hailo), has revised his guidance to startups about how much they should commit to finding a good domain. In today's market, he thinks it's appropriate to fork out up to $50,000.[1]

Concrete Branding Strategies

Once you've settled on a name, it's time to start thinking about details such as your strapline, logo, colour scheme, tone of voice and style, which, when wrapped together, deliver what is called your brand.

There's no single tried and proven way to get this in place, and every company approaches it differently. The simplest approach is to get a basic logo and colour scheme in place, and then work through all the additional elements as needed, which are likely to change a lot in the early days. The website 99designs.com is a great place to start. It is the biggest[2] marketplace for logos and identities – at a pretty reasonable price, too. You can browse loads of designs, and then launch your own competition to have designers craft you a logo and visual identity for a few hundred dollars.

Another approach is engaging a branding agency – but you might as well go Internet dating. Finding an agency that will provide you with something you're happy with is pure luck. I've run a creative process a number of times in the hope that an agency will deliver something killer. And while it's possible – Albion in London delivered the logo and branding for Skype (and I think it's pretty killer) – it will cost you an arm and a leg. You won't get much change out of $25,000 – and that's just for a logo, visual design and a branding presentation. Not great when you're bootstrapping.

I don't think there is a short cut to coming up with a good name and developing that into a great brand. It takes a lot of time – and during the early days it is an evolving process. That said, it does need constant attention and it's possible to achieve great results.

Branding stories

SQUARE. This payment app was originally called Squirrel. On the road trip I mentioned earlier, Jack Dorsey got thinking. Squirrels dart around and collect acorns and those acorns serve as a type of currency for them. And, in keeping with one of the 'ideal requirements' for a company name, the word 'squirrel' can also be used as a verb: people often *squirrel* away their money for a rainy day. Dorsey envisioned an acorn-shaped piece of hardware that would plug into the earphone jack of an iPhone, which would serve as the swipe mechanism to read a credit card. He also wanted to design it so that the sound when a card was swiped resembled the squeak of a squirrel. Unfortunately, they hit on a not-so-small problem: there was already a payment system out there named Squirrel. Undeterred, they grabbed a dictionary and started scanning. Eventually they found a word that they thought would work: a square is a fundamental shape that suggests heft. A square deal is a fair one. And, when two parties settle a deal, they square up. They were able to purchase squareup.com without too much negotiation.

HAILO. The Hailo name and brand were created by a former agency branding man who had gone independent, called Daren Cook. Hailo was the winner of a small beauty parade. The options included Rita (after the Beatles' song 'Lovely Rita'), QuiCar and Fleetr. The working name was Cabitnow, before Hailo won out. The concept was very much around the app delivering 'taxi heaven' – a great and almost religious experience. From the get-go we wanted to make sure that (a) the name of the company could be used as a verb, (b) the colours were reminiscent of taxis internationally (hence the yellow and black – we used a yellow very close to that of NYC taxis), and (c) the logo was very mobile-centric – hence why it's always been in the format of an app icon. What is less well known is that the layout of the logo is meant to be more personal because it looks a bit like a face – with the 'O' serving as the mouth and the 'TM' being a beauty spot on the bottom right.

COLOR.COM. This story is a combination of branding and product train wreck (I talk a lot more about Color later in the book). It's a doozy. You might recall that, in 2011, an iPhone social photo-sharing app called Color launched with much fanfare. Before it even launched it had raised $35 million in funding – which is, frankly, ludicrous. The company paid more than $1 million to acquire both color.com and colour.com. Definitely a wise investment for a nascent, pre-launch company! In any case, I am sure it would have been OK should the app have been successful. It turned out that the app's proposition was somewhat complex: it was about social photo sharing but with random people in your vicinity. First, the app just confused users, because its interface was bizarrely unintuitive; on top of that the technology just didn't work. The result was expensive confusion.

In the end the company folded after a well-publicised public argument between the founder, employees and investors. Just remember that a great domain and URL doesn't make an app successful! Nor does $35 million!

Taking It All Online

So now it feels as if it's coming together – name, logo, a bit of a colour scheme and a pretty decent domain name. Now it's time to announce to the world that you've arrived! And that means getting your website up and running.

Once again, you want to be efficient and not invest too much time and effort. The very first thing to do is to get a holding page up. Launchrock.com offers a very handy service here: it hosts a basic website for you; you can customise it with your logo and colour scheme with zero coding skills; you can add details about what your app does, when it's launching and how to get in touch with you.

Most importantly, and its main feature, it prompts everyone who visits your site to register their email address and get early access to your app. It then encourages those visitors to share the site with friends. Why is this great? Well, over the course of the next few weeks and months as you build your app – reaching out to designers, contractors, developers and investors – people will invariably visit your website. With launchrock.com in place you're now capturing hundreds of the email addresses of people interested in your app. That means you're building a list of potential users. For a small amount of effort you've started to build your user base.

It's also a good idea to register all of your social-media accounts – claim your Twitter and Instagram handle, your Facebook page (along with a custom URL), your Google+ and Pinterest pages. Start building a consistent online identity.

With these core elements in place, you're now ready to get down to the process of designing and building your actual app. There are a few things to consider upfront, to ensure that you're going about the process as efficiently as possible and leveraging the techniques used by the very top startups.

Chapter 7

Getting Lean and Mean

'In 2012 there were about 4,000 unique devices running Android; in 2013 it was around 12,000. About 600 different companies manufactured those devices'
#BILLIONDOLLARAPP

Too many startups begin with an idea for an app that they *think* people want. They launch into the process of designing and then building their app, often working on it for months, developing feature after feature. When the product is complete, they launch with a big fanfare and realise that users are just not continuing to download it and few people are continuing to use it. What went wrong?

Most people often don't spend anywhere near enough time talking to their prospective customers. Understanding your target users is critical – especially understanding their problems and how your app is going to solve those problems.

I recommend a great book by Eric Ries called *The Lean Startup*. 'Lean' is a great adjective. It is about building your app wisely, frugally, without wasting time, without excessive costs, and maintaining a vigour and energy. In the book Ries summarises an approach to eliminate uncertainty, and inject process and rigour around developing and testing your product to make sure it resonates with your target users. Throughout this book, I'll echo a lot of best practice Ries writes about – my goal is not to apply rules dogmatically, but rather to help you invoke your best judgement in any situation. I'll show you how to work smarter, not harder. We'll explore how to create a prototype app faster, so that you can get it into the hands of users quickly – and get the all-important feedback about whether your product is heading in the right direction.

We'll incorporate Ries's ideas of validated learning when we look at the way to measure, with analytics, how and why people use your app. And, most importantly, you'll see why the build–measure–learn cycle is so important as you quickly put your prototype together, measure feedback, make tweaks and then try new features and improvements with your users.

Building apps is a very fun process – and one that is constantly becoming more accessible and more simple. Let's start the journey by tackling our first major decision.

To iOS or to Android? That is the Question

You are now *almost* ready to start building the first version of your app. There is one more decision to make before you get going: which platform you build your app for first. This decision is important for a number of reasons.

One is that iOS and Android users are actually quite different – in terms of likelihood to spend money on their smartphone, in terms of demographics and even in terms of geography. Another important reason is that the design pattern – namely, how your app is designed and interacts – is quite different between the two platforms (this determines the designers and software developers you recruit first). And perhaps the most important reason is that, at this point, you need to stay focused – don't double the effort (and cost) on Day One by trying to support two operating systems.

A great approach is to focus on one platform and make it a success. Instagram hit the billion-dollar mark with just an iOS version of its app (though the successful launch of its Android app did help seal with the deal with Facebook). Similarly, Snapchat, WhatsApp, Hailo and Angry Birds all launched on iOS first. They then focused maniacally on making one platform successful – before undertaking the additional effort and cost of supporting a second one.

It is worth remembering that each platform has an enormous and very loyal user base. That means users are accustomed to the specific way their 'preferred' platform works – and that means you need to respect and understand those specificities to win over the users of each platform.

Irrespective of which platform you focus on first, you should become an expert on both platforms, so carry both an Android and an iPhone. Use them both religiously. You'll become accustomed to both quickly, and start to appreciate how different they are. Armed with this first-hand knowledge, you will create a better product – a better app – for your users. And, for a founder, that is one of the key responsibilities.

One at a time, please – but which?

So how do you decide between the two platforms? Let's have a look at the decisions made by our model citizens – the Billion-Dollar App Club.

WhatsApp decided to launch on iOS first. Its logic was that iPhone users also disliked mobile advertising – and didn't want their app experience to be junked up with intrusive ads. WhatsApp also reasoned that iPhone users spend more on premium apps – in fact they were spending 2.7 times more than Android users.[1] Apple users accounted for 73 per cent of the total spent on the App Store and Google Play as of mid-2013. The

WhatsApp go-to-market business model was to charge users $0.99 for the download and offer free messaging for ever.

It worked: iPhone users were happy to pay $0.99 for the download – and did so in the tens of millions. With big revenues rolling through the door, WhatsApp was ready to launch an Android app. WhatsApp launched on Android with precisely the same business model: a $0.99 download and free messaging for ever. In a complete surprise, download numbers struggled to such an extent that the team ended up quickly ditching the $0.99 price point for Android users, and made the app free. Downloads rocketed, the user base continued to grow and amazed iPhone users were still happy to keep paying $0.99 to download the app.

For almost precisely the same reasons, Rovio released its Angry Birds game exclusively for the iPhone in December 2009. It offered both a free version with a limited number of levels, as well as a premium version where you could pay $0.99 for more levels. Its focus was entirely on delivering the best experience to users on that platform, and, as a result, it built up huge demand for the game on Android. It launched it almost a year later and by December of 2010 its CEO announced Rovio had hit 42 million downloads[2] – 12 million of which were paid, and on iOS.

What about iPads?

Supercell – creators of Clash of Clans – actually launched their company with an all-platform strategy. But what Ilkka Paananen, CEO of Supercell, and his team realised ever so quickly was that, if you don't build your games from the ground up for a specific platform, you're not going to build the best games:

'We started from this online web product, and how we actually discovered tablet was when we started to create a version of the game for the tablet, and we realised, "Hey, it's not going to be a good game!" Unless we actually start from the tablet, we're never going to create the best games for this platform.'[3]

When billion-dollar app Flipboard launched its business, it completely ignored a smartphone app. Instead, it focused 100 per cent on building an iPad app. Why is that? Flipboard's mission was to create the best 'social magazine'. The magazine experience was a lot better on a tablet format, and the best tablet was the iPad. When Flipboard launched in late 2010 the iPad was only seven months old, and Android tablets were struggling to compete in any meaningful way. So this platform decision was driven by a unique and superior user experience – and a big gamble that an entire user base would materialise for the new mobile device.

The flipside of taking such a gamble – and building such a great app – was that Apple was there to support Flipboard all the way, so much so that Apple named Flipboard its iPad app of the year in 2010[4] – an announcement that catapulted the fledging startup into the stratosphere.

Getting Android right

One of the big reasons that you'd want to potentially launch your app on Android is to reach a larger audience. The Android operating systems were expected to be running on more than 1.9 billion devices by the end of 2014, compared with about 700 million for Apple's iOS operating system.[5] Android is picking up especially well across Asia and Africa, due to its cost advantage (the operating system is free for hardware manufacturers and mobile operators to use). So, if your app is focused on users in those regions, then Android might be the first choice.

Whichever platform you start with, it's worth keeping in mind that, if you're truly dedicated to building a billion-dollar app, you're unlikely to do that by mastering a single platform, so you will need to build a version for both eventually. We learned this the hard way when I was at Hailo.

When we released the first version of the Android passenger app, we basically just reused the design we had for the iOS app, hoping to save time and get it out there. The feedback we received on Google Play – and directly from users – was overwhelming: 'This is just a repackaged iOS app!' We didn't win many fans, and, for the first time since the launch, began to receive lukewarm reviews.

One of our designers, Dave Clements, made it his personal mission to 'Androidify' the app himself (spending weeks of late nights delivering it). When the new version went live, the feedback was so positive that even Eric Schmidt, the CEO of Google at the time, and Hugo Barra, the VP of Product Management at Google, made a point to personally get in touch with the team to congratulate us on the improvements made.

Other platform considerations

When you launch your app, you need to test it out on the smart-phones that will run it.

This is pretty easy on the iPhone, since it comes in only two screen sizes – iPhone and iPad. This means you need only two devices to test your app on. Apple has kept it simple, because it builds both the phone itself and the operating-system software that runs on the phone. If you want to be thorough you can test your app on older hardware such as the iPhone 4 or 4s as well.

Android, on the other hand, is more complicated to get right. This is because it is open-source and has therefore been adopted by hundreds of companies to run their smartphones – so you're look-ing at a slightly bigger selection of handsets to test on.

In 2012 there were about 4,000 unique devices running Android; in 2013 it was around 12,000. About 600 different companies man-ufactured those devices.[6]

Hmm. Not sure you can fit that many handsets in the office.

What does that mean? It means if you focus on Android first, you'll need to support a lot more than two screen sizes from the start. It also means that everyone needs to think about screen size all the time – from your product managers, to your designers, to your engineers. It also means you need people with rich experience of working with Android – you don't want to have an inexperienced designer cut their teeth on your Android design.

It also means you need to have at least 10 of the top Android handsets in your office so that you can test that your app works on them as expected. And, yes, there are big differences between how your app can or might behave on an HTC smartphone versus a Samsung one. If you don't test thoroughly, you risk annoying users unnecessarily. And it may not just be something superficial that goes wrong: it could be something a lot more operational. We saw a wide range of issues with the Hailo app on Android. Supporting Android added significant overhead.

Another consideration is which *versions* of the operating systems your app supports. When we talk about these two operating systems – iOS and Android – it's worth noting that they both have a substantial update about once a year. It can be the equivalent of upgrading your desktop from Windows 95 to Windows NT to Windows 8. There is a good chance that your app will require development work to make it function perfectly – or at least a lot of testing.

Recent changes have made updating the OS a bit easier – in late 2011 Apple revamped its operating system to update itself over the air. This is what happened with the latest upgrade from iOS Version 6 to Version 7. Within a few weeks of its release, about 75 per cent of all Apple devices (that's about 230 million) were running the new version. [7]

In early 2014, there were 600 companies running a version of Android on their smartphones. It means there are eight different versions of Android running in the wild.[8] In reality, it's not quite that bad: the latest three versions of Android account for the vast majority of devices. But it does mean you have to make some hard decisions about which of your users to support – and which ones you need to let down.

Chapter 8

App Version 0.1

You've made your decision to build on iOS or Android. Now you need to get your head around what you're actually going to build.

Everything starts with Version 0.1 – it's the very first iteration, the prototype, of your app (Version 1.0 is reserved as the first version shipped to the public). In this version you want to focus on the most basic set of features that will make your app unique, useful and different. It's often called the MVP (or the minimum viable product).

At this point you want to focus on only the parts that are absolutely necessary to show why your app delivers something new and novel – something that wows your users. For Hailo it was focusing on how a user could see nearby taxis on a map, then hit the 'Pick Me Up Here' button and have a driver accept the hail. We also added in the ability to see the driver come towards you. That was enough to make people feel a wow moment.

Since we had limited time and resources we ignored everything else – there wasn't a proper way to register an account, you couldn't add your name, you couldn't interact with the map, and you could see only the taxi's number plate. It was bare bones. But it delivered a feature that no one else could offer!

But it wasn't quite that simple with Hailo. We needed to actually validate two different propositions, with two different apps – one for drivers and one for passengers. All on a tiny budget.

Faking It

When I joined Hailo my first task was to figure out precisely what the prototype app needed to be. We had a small amount of funding that was keeping the lights on. We had a very short amount of time to design and build an app to demonstrate our vision – and then objectively validate with users out in the real world.

The first decision was relatively simple: should we design the passenger app or the driver app first?

We knew that, if we couldn't generate enough interest from drivers to use the app, there wouldn't be a large enough supply of taxis, which would mean no passengers using the service, which would then mean we didn't have a business. The passenger app could clearly wait.

Great. So Jay, Caspar and our taxi-driver cofounders – TRG (Terry, Russell and Gary) – were all full of great ideas about what features to include in the driver app to make it appeal to drivers. After a long debate we had agreed on three features – and they were baked into the very first 'paper sketches' of the driver app. Now came the long – and iterative – process of actually transforming those sketches into a piece of software.

Every week I hosted a group of about 15 taxi drivers for breakfast at a café and restaurant called Smiths of Smithfields. They were all keen to give us a round of colourful weekly feedback: 'These buttons are way too small, gov,' said one driver. 'I think your fingers are just too fat,' replied another driver. I couldn't disagree – he did seem to have pretty fat fingers.

Initially, we used paper designs of the app to feed back about the proposed features, then iterated designs, received more feedback and delivered the basic app built by our development team (at the time just three guys).

The first three features that we included were the ability for drivers to report a 'burst' (somewhere that needed more drivers, such as a concert that just finished, ejecting concertgoers needing cabs in their hundreds, or a taxi rank that had not enough taxis), automatic statistics about how far they had driven during a shift and what percentage of their time they were occupied, and the ability to process a credit card by entering in the card details into the driver app.

We managed to cram all the features from our drivers' feedback into that early version of the driver app. Actually, we cut one big corner. At the time, we couldn't actually process credit cards with the app – we just faked it. Since we were working with a trusted group of drivers, we just recorded every attempt to pay by credit card (customers weren't charged).

The drivers were impressed, and our test group started using the app quite frequently. They communicated with each other about where there was demand for taxis. At that point we needed to get proper credit-card-processing services integrated (it was pretty quick to implement). Now drivers were able to process credit cards from passengers who requested it, and Hailo started to generate real revenue. That was a great milestone.

The driver app was still prone to crashing, didn't have a registration process and had a simple but not very attractive user interface, but it worked. And it delivered features that drivers could not get anywhere else. It was a win.

So at that point we knew we were doing something right. Now, we had to close the loop and get passengers on board.

Getting to the First Wow

The big challenge that we set for ourselves was having a user try to hail a taxi with the app, and then the taxi actually arriving. We knew that no one else could deliver this kind of experience. We tested the concept with people at a nearby Starbucks by walking them through the app experience on paper. We knew that, if we could deliver on that promise, we could get to 'wow'.

Here's what the experience looked like:

- Open the app and we'll show you taxis near you on a simple map
- The app will tell you how far away in minutes the nearest taxi is
- You can press one big yellow button, 'Pick Me Up Here'
- Once you press the button, we contact drivers
- A driver replies, and the user gets a confirmation when the driver accepts
- The user sees the driver arrive on the map – with their registration plate number.

That was the bare-bones prototype we wanted to build. We were able to build it in a matter of weeks, cutting plenty of corners along the way. We even added the ability for the customer to pay with a stored credit card (yes, we faked it again).

The first time I summoned one of our test group drivers he actually drove quite a few miles to pick me up, and I have to say I that was truly wowed. I couldn't believe it had worked.

Really Viable

Building your own prototype is a tricky and iterative process. What you are trying to do is create the bare bones of something – the very basic vision of your app – and see whether it can become something that people love. You need to get to wow as quickly, cheaply and efficiently as possible. There's no point wasting time or money on any app that doesn't get to wow.

The point of doing it this way – using paper designs, testing the bare minimum – is to get real data, to get real validation. Imagine the opposite: spending three months designing something in detail and another couple of months developing the software without any feedback from users. Unless you're testing absolutely everything with your target users, you have no idea whether it will work.

Hailo required a rather involved prototype because we required two apps, and they needed to talk to each other. But we still cut plenty of corners to seek the data we needed. And this was very much a prototype and definitely not something we could release to the public.

Seeing it in action prompted our existing investors to continue supporting us, and it persuaded a new investor to back us and get us to the point where we had a proper Version 1.0 product. You're constantly attempting to validate your big idea – with real data, real user reactions and feedback.

Wireframes and User Journeys

During this developing-and-testing process, we actually went through another step. To visualise the prototype you want to build, you actually need to start to put together an increasingly detailed blueprint of your app.

This blueprint has two goals: (i) to illustrate what each screen of your app looks like; and (ii) to explain how your app behaves. Since smartphones are small, intimate and all based on touch, there are lots of ways you can make your app behave.

Together, these elements are called the 'wireframes' for your app. They are created and owned by the person or team in charge of the product development. They are used as the blueprints to communicate everything about your app. You'll use them to explain how the app works to potential users (in a Starbucks test or similar), and you'll use them to communicate to your engineers how to build your app. Other people involved in the app-development process, such as testers, will also need to understand these documents. Their job will be to compare and test what the app is supposed (designed) to do compared with what your developers have actually implemented.

There are many great tools to help you do this. Moqups.com is a simple and great one. There is another one called Balsamiq.com, and then a more advanced one called OmniGraffle. Some people prefer jumping into Photoshop at this stage. But use whichever tool you're most comfortable with.

If you're a complete first-timer, I'd suggest using PowerPoint to create a slide-by-slide outline (using the shapes elements) for every screen in your app. Immediately, you will see the challenge of translating the idea of the app you have in your head to precisely how it's going to look on a screen. It's challenging.

Once you've got a comprehensive wireframe outlined, you'll be ready to work with a designer. Your wireframes will provide the basis for a structured conversation and the scope and size of the app that you'd like to build, and a designer will be able to create a pixel-perfect design.

Great Design

> Good Design is as little design as possible – less, but
> better – because it concentrates on the essential aspects,
> and the products are not burdened with non-essentials.
> Back to purity, back to simplicity.
>
> *– Dieter Rams*

Dieter Rams is one of great pioneers of industrial design. For
decades he worked at Braun and pioneered state-of-the-art radios,
audio equipment, cameras and furniture. He has been exalted by
many as the leader of 'minimalist, intuitive design'. Apple's lead
designer, Jony Ive, is one of many who have been massively influenced by his style.[1]

Rams is celebrated for his 10 principles of good design[2] – something that is critical today. Keep these principles in mind as you
design your app.

According to Rams good design:

- Is innovative
- Makes a product useful
- Is aesthetic
- Makes a product understandable
- Is unobtrusive
- Is honest
- Is long-lasting
- Is thorough down to the last detail
- Is environmentally friendly
- Has as little design as possible.

Design matters because competition in the app world is heating up and because people can be fickle. Twenty-six per cent of users will open your app once and never use it again.[3] From that very first use you need to be able to deliver value to a user; you need to make them smile; you need them to say, 'Wow, this is really cool!'; you need to set an expectation and deliver.

You're still at a stage where every dollar counts, so you need to find a way to get your design work done as quickly as possible, for as little money as possible (simultaneously, you want to be grooming your designer to join you full time when you get funding in place).

Your goal is to get a designer to translate your wireframes into pixel-perfect mockups of your app. That basically means a set of screenshots and files that will look the same – pixel for pixel – as each screen of your app. Once those files are prepared, it is relatively simple work passing them on to your developers to implement as software code.

One good way to expedite the design process is to become a bit of an expert in what constitutes a great app design (that's what great product people do). Even if you aren't a natural at design, that doesn't mean you can't teach yourself what works. Mobile-app design, I find, is a lot easier than website design because it's a lot simpler. It is straightforward to tell if something is functional on mobile. It's also very easy to get lots of opinions quickly by sharing the app – or just the screenshots of your app – with anyone who'll listen. Ask pointed questions about specific things, and record all the feedback you get.

LEARN FROM THE BEST. A good place to start is to go out and download the top apps in each app-store category. It's not by chance that millions of people regularly use these apps – and spend hundreds of millions of dollars through them. A large part of their ongoing success is because of their design. So download all the current billion-dollar apps – Snapchat, Flipboard, Angry Birds, Puzzle and Dragons, Uber, Candy Crush Saga, Instagram, Square, Waze, WhatsApp, Viber, Tango, Pandora and, of course, Hailo. Make sure you're an expert with other leading apps such as Facebook, Dropbox, Evernote, Amazon, eBay.

DISCOVER AND COLLECT SCREENSHOTS. It's helpful to build a collection of screenshots of your favourite apps and favourite features. It's your personal library of great designs that you can access at the tap of a button. This ends up being a great resource – and reference library – when you want to build a new feature. A good idea is to store everything on Dropbox: its mobile app is great because it allows you to browse through galleries of screenshots very easily, and, of course, it's super easy to share them as well.

A great source of inspiration is www.pttrns.com – it's home to thousands of screenshots. They are wonderfully arranged by 'feature' – such as how to create an account, play music, swipe through photos, edit personal information, navigate a map. You can compare instantly across super-successful and freshly launched apps. But be careful with what you use as a benchmark: just because they appear on this or that site, it doesn't mean they're effective.

FUNCTIONAL VERSUS BEAUTIFUL. There is a fine line between beautiful and functional design. The highest praise is reserved for apps that achieve both. But let's be 100 per cent clear: functionality should be your number-one priority. WhatsApp is arguably a rather spartan app, and not super-pretty, but it's damn intuitive, has great performance and always works. Similarly, Snapchat has a simple and uncluttered interface, and, despite requiring you to learn a new behaviour to view a snap (press and hold to view the content while the clock counts down), achieves that goal easily.

Hunting Designers

So you have your wireframes sorted out, you know you want a well-designed app, but where do you find a great designer?

Designers are elusive creatures who can be hard to recruit at the best of times. The great ones are always booked months in advance; most don't want a permanent job (they're happy doing contract work – going from new project to newer project); and, when you do find someone, you usually find that a combination of very strong design opinions and outrageous day rates will make it hard to get the designs you need.

A great approach is to trawl two websites in particular – dribbble.com and behance.net. I've experienced great success with Dribbble (yes, three Bs in its name and in its URL). In fact, that's where we found Hailo's head of design. The great thing about these portfolio sites is that you can see a designer's style, often with a lot of their historic work. Also, Dribbble has added the ability to search for designers who are actively looking for full- and part-time work (and, despite its being a paid-for feature, I thoroughly recommend using it).

Once again, I would also suggest checking out AngelList: it attracts the most entrepreneurial cross-section of people, across all areas, whether in design or engineering.

Rapid-Design Prototyping

You've created some detailed wireframes; you've refined them through feedback from your target users; and now you're working closely with your designer and developers. You're making progress – and the app seems to be coming together.

A number of increasingly simple and yet powerful tools have

come onto the market that allow you to make visual designs more interactive than ever before. A great example is proto.io. Hailo's designers used it quite a bit. It gives you the ability – both on the desktop and via an app – to very quickly and cleverly link together screenshots (even rough, quick ones) in a way that makes them clickable – and linked to other screens.

Why is this important? A huge part of app design is the inter-action – screens are small, and you need to use the 'real estate' wisely. Often, less is more. You want to communicate clearly what is happening on a given screen – and you want clear calls to action. You want to remove or clarify anything that is confusing.

Tools such as proto.io allow you to put a 'prototype' app directly into your hands – and you can instantly see how it feels and behaves. You can test your ideas without the need to bother with developers – and get meaningful feedback. No software development involved.

Get Coding

So now you're ready to start writing some software for develop-ing your app (if you haven't already). Remember, there is no single correct way to build your app, but I'll outline all the options avail-able to you and highlight the proven, best strategies – ones that have led to blockbuster apps.

You're probably in one of two camps right now: either you are developing the app yourself (or your cofounder is), or you're out-sourcing the development of your app to a contractor (hopefully, someone really talented and not too expensive) – though you need to be thinking about how to get a developer to join your team.

If you invested time and energy creating some thoughtful wire-frames, benchmarked the key parts of *your* app against some leading apps, hunted down a great designer and then found a solid developer to translate all of that into software, then you're going to have a very solid app by the end of this chapter.

Don't worry about having it super-polished – what you really want to be able to do is put it in front of target users. Your goal is to get your gaming app in front of gamers, to get your taxi app in the hands of people who take taxis frequently, to get your payment-processing app in the hands of merchants who need it. This is about testing your app solution in the wild – with actual, real potential users. You should be shooting to get it into the hands of at least a few hundred people, ideally up to a thousand or so.

Measuring Good

There's one more thing you need to think about before you start coding: making it super-easy for users to send you feedback. In these early stages, don't make a user dig around your app to find a feedback form: make sure it's front and centre and makes it easy to get in touch with you with feedback. Reactions to the early versions of your app are key to success, and you want to use real user feedback to tune the direction in which you're heading.

I would love to think that all the people using your app would be happy to send lots of feedback all the time, but the reality is that you hear only about the extremes: the great experiences and the very bad experiences.

But, to make your app a success, you need a systematic way to understand the experiences of all your users, what they are actually doing on your app, and what they like and dislike about it. How do you keep your finger on the pulse and figure out what's going on?

Analytics is the answer. You can include snippets of analytics software in your app so that you can automatically track what every single user is doing on your app – what they are looking at, what they are clicking, how long they spend performing an action and at what point they open and close the app.

Analytics tools give you a powerful visualisation of this information on both an individual user level and a level that shows you what the population of users is doing on your app. How you interpret this information, i.e. what insight you can glean from it, and what you should do as a result – well, that is an art!

One of my favourite solutions, which is also one of the new market leaders, is mixpanel.com. This tool gives you a powerful way to view your users on an individual and group level, allowing you to see precisely what they are doing every time they fire up your app.

There are plenty of solutions out there for analytics – and it can be a bit overwhelming selecting one. Starting off, I would recommend putting in two solutions: a robust free solution such as Google Analytics (which has been improving in leaps and bounds on the mobile side) and something like Mixpanel, because of its powerful communication capabilities and its very powerful ability to segment and analyse your users. We'll delve more into that in the next chapter.

Other solid solutions are Flurry (it's free, has probably the biggest footprint, but lacks powerful segmentation and its geographical filtering is weak), Localytics is very good (though it doesn't have as wide or robust a feature set as Mixpanel), and then Kontagent (we found it to be quite a bit behind the other solutions, and focused a lot more on sales rather than product or engineering).

Chapter 9

Metrics to Live and Die By

Successful Internet and app entrepreneurs are obsessed with metrics. If you can't measure something, you can't improve it. From the start, you need to both focus on the correct metrics and make sure that you have the analytics in place to measure them reliably.

Let's start with a simple and powerful framework made popular by the entrepreneur and super-angel investor behind the 500 Startups accelerator and Seed Fund, Dave McClure (he was also the director of marketing at PayPal for three years).

There are five types of metrics to remember (they are quite pirate-like – AARRR):

- **ACQUISITION**: users downloading your app from a variety of channels;
- **ACTIVATION**: users enjoying their first 'happy' experience on your app;
- **RETENTION**: users coming back and using your app multiple times;

- **REFERRAL**: users loving your app so much they refer others to download it;
- **REVENUE**: users completing actions on your app that you're able to monetise.

You're going to need to define specific metrics in each one of these five categories as soon as possible; some of them are going to be specific to your app, your market and your business model, while others are going to be pretty similar no matter what type of app you're building.

One thing to remember is that all your metrics should be valuable and actionable, and they shouldn't just be vanity metrics. Valuable metrics are ones that drive decisions.

All these are best explained with concrete examples, so let's look at how Hailo did it.

ACQUISITION. Acquisition isn't just getting a download. It's about acquiring a user – so you need to set the bar higher. A user should be someone who downloads your app, opens it and clicks at least one button, or has a session length of more than 10 seconds. Counting someone who abandons your app as an acquisition isn't particularly useful (and nor is the channel that brought them to your app).

ACTIVATION. This is about a minimum threshold of engagement – someone who has completed an action that might lead to potential revenue. At Hailo we counted a user who created an account as an 'activation'. Alternatively, you could count someone who has clicked X times within your app, or someone who has a session time of more than 60 seconds.

RETENTION. Understanding and driving this metric is critical to success. If you can't keep users coming back, then your app is doomed. At Hailo we measured this as how many times a user opened up the app per month, and how many times they opened the app *and* requested a taxi. You can also drill down to the level of measuring how many times per month a user opens or clicks on the emails that you send to them.

REFERRAL. This can be a tricky metric to track, so it helps to build a product feature to encourage it. From Day One we built in the ability to input into the passenger app on Hailo 'promotion codes' that would give passengers £5, £10 or more off their next taxi ride. The system was flexible enough to track every single 'referral code' back to a user, a driver or a specific marketing promotion.

REVENUE. You should be aiming to make this work from the very beginning. We monitored not only spending per customer, but also any discounts, refunds and the actual gross margin we made per user. We also monitored how much people were tipping their drivers.

Putting Metrics into Action

Your goal – even at this MVP stage – is to put these metrics into action. You don't need to delve into more complex ones at this stage, but you do need to understand how these metrics map out your user lifecycle.

The first thing you want to do is plot out what you think your user lifecycle is going to be. This is best explained with a real example – let's use Hailo again. Our first model was pretty simple (and was refined down the line).

Stage	Conversion Goal	Notes
Acquisition	User downloads app	Measured, but not used as a measure of success
Acquisition	User launches app, either taps any button or session time greater than 10 seconds	This was Hailo's 'higher bar' version of acquisition
Activation	User creates an account	A user who has shown interest in becoming a passenger
Activation	User adds credit-card details (required for fare payment)	A user who has shown real interest in becoming a passenger
Activation	User requests a taxi	Hailo still has to provide user with a driver
Activation	User receives confirmed driver	Hailo's job is done – now it's up to the driver
Activation	User gets into a taxi	As good as revenue (but not just yet)
Revenue	User completes ride and pays	The Holy Grail moment

At the beginning, Hailo focused the user lifecycle very much on a single taxi journey. It was only later that we tuned our lifecycle to include the retention and referral metrics as well.

Your goal will be to focus on as many of those five metrics as possible – and then, throughout the next few weeks and months, test your app, measure the results, and iterate to improve the features and try to move your metrics in the right direction. Specifically, you'll want to focus on improving the conversion of downloads to 'acquired users', the conversion of 'acquired users' to 'activated users' and then finally converting your 'activated users' into 'revenue-generating users'.

Chapter 10

Let's Get Some Users

I bring this up now not because you should be focusing on actually getting users to download your app at this point, but because you need to understand how you *will* get users to download your app in the future. You need a viable – and realistic – plan to drive downloads, installs and regular usage.

If you're new to the app game you'll underestimate this. Even if you're in the mobile world, unless you actually have to work with marketing you will also probably massively underestimate this as well.

The competition is continually heating up. There are hundreds of thousands of apps on the Apple App Store alone – which means you are competing in a very crowded space. That translates to higher advertising costs – something Google, Apple and Facebook are quietly very happy about.

This section will give you a rundown of how user acquisition works – in broad brushstrokes – so you're fully aware of all the elements available in your toolkit and the likely costs you will encounter (or need to invest in) further down the line in order to drive the downloads and usage that you need in order to make your app successful.

Later chapters will dive deeper into mastering all of these components – so right now focus on understanding the basics of app distribution.

How Will Users Find Your App?

That's the magic question. You can have the best idea in the world, but it means nothing if nobody can find your app and start using it. To answer, let's have a look at the research. The data says that organic searches – meaning people going to an app store and searching for an app there via, say, the Apple App Store or Google Play – count for the vast majority of downloads at 63 per cent,[1] while 61 per cent[2] of users find an app via the search box in the store.

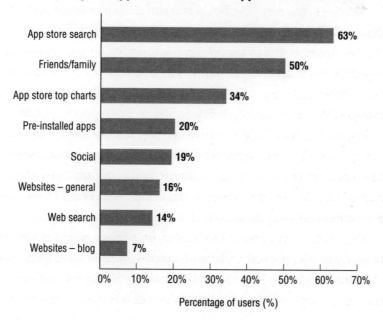

Percentage of App Users Who Find Apps Via Each Channel

According to Ankit Jain, Google Play's head of search and discovery, 'For the average app, search makes up the majority of installs',[3] and 12 per cent of daily active users (DAUs) on Google Play search for new apps daily, with a further 50 per cent searching for apps weekly. This explains how significant search is within the app store as a driver of downloads – you don't want to ignore that.

How to be Found in the App Store

It's a bit of a new-fangled term, 'app-store optimisation' (ASO for short). Given that there are hundreds of thousands of apps on the Apple App Store alone – as we saw earlier – it's not as easy as it once was to find an app. So that means you need to make sure you do everything necessary to make *your* app as 'findable' as possible. ASO is equivalent to search-engine optimisation (SEO) for websites (which I talk about below).

Here are some concrete things you can do to maximise your app 'findability'. I'll start with the things you can directly control, and then go through things that you can influence, but not necessarily control 100 per cent.

YOUR APP TITLE. Your app's title is the most important factor in achieving a good ranking in an app store. It's like the '<title>' tag in your HTML (which contains keywords to tell people and search engines what your web page is about), and communicates to both users and app stores what your app does. You want to include both what your app does (as well as keywords) and branding, i.e. your name. Always make sure it looks user-friendly and is legible. Make sure it parses well and makes sense. Examples are 'WhatsApp Messenger', 'Square Register', 'Square Wallet'.

DESCRIPTION. You see this field only in Google Play. Patrick Haig, vice president of customer success at Moz.com, likes to break descriptions down into two sections: 'above the fold' and 'below the fold' (these terms refer to the content that is immediately visible on your smartphone screen before you have to scroll down any further). He says, 'Above-the-fold language should be 1–2 sentences describing the app and its primary-use case, and below the fold should have a clear and engaging feature set and social proof.' An example would be, 'Open Square Wallet to find great local businesses anywhere you are. Square Wallet puts your credit card, loyalty cards and receipts into one app.' You can then add secondary information describing your app in a lot more detail, including user and partner reviews, below the fold.

KEYWORDS. This field appears only in Apple's App Store (it is the equivalent of the 'description' field in Google Play). You have a 100-character field to tell iTunes which keywords your app should show up for in searches. Given that you have so few characters, you should use them wisely. Here are some tips: focus on relevant, high-volume and unique keywords; don't use strings of words or phrases, as Apple will combine individual words for you; don't repeat words already in your title; separate keywords with commas; don't use spaces (not necessary: the commas are separating your words for you, and each space, don't forget, counts as a character).

ICON. Make sure you have a clear, memorable icon that really communicates your brand. You should be very thoughtful about how it works across all channels – and not just in the App Store. Be mindful that you have space for only about eleven characters within the icon (I say about because it varies, i.e. the letter 'm' is a lot wider than 'i' for example) for your app name, otherwise it will appear truncated on a user's phone, so check it first! It's also something worth thinking about when you come up with the name for your app.

SCREENSHOTS. Both app stores allow you to upload screenshots to sell your app. These do not have to be 'screenshots' *per se* – you can tune them to communicate the details you want about your app. Make sure they are very much promotional shots of your app and that they clearly communicate core functionality and features – so it's very much worth having your designer make them as sexy as possible. This is your storefront window – so make it sell!

That's a brief overview of the basics you should know about the app stores. We'll dig into more details later in the book.

Your App's Website

With the app stores sorted, you should make sure all the other basics are covered. Even if you're an app, you need a website. If people Google you, they should be able to find you easily. Ideally, you want your site to focus on helping people download your app – i.e. a crisp, snappy description – and then a clear download button that links directly to the app stores you're on.

Naturally, your website should work on a desktop Web browser, a tablet and a smartphone. Delivering a great experience across all these devices is called 'responsive design'. A good example is whatsapp.com: the design elements on the webpage are fluid (they resize smoothly) when you look at them on a desktop browser, but, when you look at the site via a smartphone, the content is laid out in a much more mobile-friendly way, with bigger touch areas and an interface that's simpler to scan and scroll. It even resizes itself when you're browsing from a tablet. Another great example is squareup.com. Open up the site – on a smartphone and a desktop browser – and check them out side by side and see how well the experiences are optimised for each device.

Another worthwhile optimisation – this is done on hailocab.com, too – is to check the device type on which a person is browsing the site. If a user browses your website from an iPhone, then instead of showing two links – one for Google Play and one for the Apple App Store – you should drop the Google Play one. It's irrelevant to an iPhone user. All these little things make a difference, unclutter the experience and help steer the user to download your app.

If you're looking for something that will expedite building your website, you should try Bootstrap. It is an open-source, fully customisable website front end – and, yes, it is responsive. It was created by our friends at Twitter. I've implemented it a few times, and it's a real pleasure to use – and a huge timesaver. This will replace the very basic Launchrock holding-page website that you had up previously.

Website SEO

SEO – search-engine optimisation – is the art of making your website appear at the top of organic (Google or other search engines') search results. Tomes, bibles and countless doctoral papers have been written on the subject – and, no, I won't try to summarise all of them here, but let's go through making sure you have the basics sorted out.

There are a lot of resources out there, and I've shared a great collection on mybilliondollarapp.com. But here's a synthesised Top 10 list.[4]

1. DOMAIN NAMES. Protect your domain name by buying it as soon as possible along with any obvious misspellings (to capture traffic from people who type your website name directly into the web browser, rather than going through a search engine), because, when you become big, someone else will buy them. Don't spend a huge amount of money on them, but it's worth paying $10 a year for the obvious ones. Also, if your app is clearly going to have international appeal – which I hope it does – grab the country-specific suffixes such as .co.uk, .fr and all the countries you think are pertinent. Don't forget .cn (China) – these guys are pretty fast at picking up what will be a very useful domain for you in the future (and you don't want them selling it back to you for thousands of dollars down the line).

2. ALEXA. Check your alexa.com traffic rank. This free tool tracks the popularity of sites on the Internet by sharing the data of many thousands of users. Once you know your rank, you can see whether you're improving it.

3. KEYWORDS. A lot of SEO is based on keywords. So make sure that your website – all aspects of it, including the page titles, page content, URLs and even the tags of photos on the site – contains the keywords you think people will be using in Google to find your app. The most important places to put the keywords are probably the 'title tag' part of the HTML in your web pages (i.e. the words between '<title>' and '</title>') and the page header (another part in the HTML of your pages – i.e. the words between '<head>' and '</head>').

4. BACKLINKING TO YOURSELF. Another good strategy is to make sure that you link back to your own content within your website – particularly your help pages, your press pages and especially any content on your website. Don't go overboard to the point of annoying users, but remember to do this, as it makes your site more attractive to search engines.

5. CREATING A SITEMAP. This is something you do to help out the 'crawlers' – the automatic programs that Google and other search engines send out into the Internet to read your website and make a map of all the pages on it. The better your navigation – i.e. the simpler and fewer clicks it takes to get to any page – the better. And providing this as a single file gets you indexed – or read – faster by search engines, which means that any new information you publish to your website will appear in all search results more promptly.

6. IMAGE DESCRIPTIONS. The indexing programs that search engines send out can't read the text in the images on your website. So make sure that you add as much descriptive text as possible to your images. Start with what are called 'ALT tags' (used in HTML to specify alternative text to be shown when a given image can't be rendered) – including the keywords people will use to find you that we talked about above.

7. FRESH CONTENT. One of the biggest things you can do to improve how high on the search results your site appears is to make sure the content is updated regularly and often. That means pushing out fresh page content, updating press releases, posting news articles, adding new images and especially making sure that you're publishing new blog posts (ideally at least once a week).

8. MAKING IT SOCIAL. Once you have this content, you want to get it out into the world. Distributing it via your social media – Twitter, Facebook, Pinterest and Google+ accounts – is key, especially if it prompts people to link back to your website.

9. LINK, LINK, LINK. SEO is all about linkages – so feel free to reach out to other businesses (via their webmasters or connections you know) and ask them to include a link to your website on their blog or list of partners, or any other way from their website. It all helps. But definitely don't use link farms,[5] or anything that sounds too good to be true. Keep it clean and build it up over time.

10. GREAT PRESS KIT. Make sure you have at least a mini section on your website for press – and a press kit. Include all the info about your app and company: photos, logos, screenshots, 'about us' pages. The simpler you make it for journalists and bloggers to find information about your company, the more likely it is that they will write about you. Don't make it hard to find all the information that they are looking for.

How Mobile Advertising Works

Mobile advertising has come a long way very quickly. At this point in the lifecycle of your app you should invest a small amount of money in advertising to generate some downloads of Version 0.1 – and see how real users interact with your app. Your goal in this section is testing how much users like your app on a small but meaningful scale.

The magic term in mobile advertising is CAC, or the Customer Acquisition Cost, i.e. the price you need to pay for someone to download and install your app. It's also called the CPA, the Cost Per Acquisition, or the CPD, the Cost Per Download. It's a great concept because it allows you to figure out – very simply and quickly – whether you have a business: if a user is generating more revenue than it cost you to get them to download and use your app, then you're on the path to profit! If, however, you're just spending money, and the users aren't helping you generate revenue, you're going to run out of money, so this is what you should be looking at closely as soon as you start to run your advertising campaign.

For example, say your average CAC is £1 and you're charging a user £2 for a download – good news! You have a profitable business (so long as users keep doing this in droves). But, if your CAC is £1.50, and your app is free, and you're counting on people to buy virtual goods, for example, and they spend only £1 on average over a six-month period, then you're going to have a challenge ahead (i.e. you'll need to spend a bunch of money upfront, and hope that over 12 months they will spend more than £1.50 with you – or a *lot* more than £1.50 if you really want to succeed).

One of the biggest and easiest-to-use platforms is Google's AdMob. It is the indisputable number-one player in mobile advertising. Google was already the largest online advertising company when it acquired AdMob, itself the leader in mobile ads, in 2010. It is now integrated into Google's AdSense platform. It's very simple to use: create an account, create a simple ad and then you're ready to go.

The way AdMob works is by trying to adhere to your desired CAC – as an *average* over time – so over the course of a week you may pay a bit more for some downloads, and a bit less for others. The goal is to hit your desired CAC.

AdMob's platform also optimises which apps and mobile websites your ad (e.g. an ad to download your new app) appears, taking into account how many people click your ad and historic performance, but also such things as the type of phone, the country and language and numerous other parameters.

You can set a budget for the day or week, and then see how those users from those specific marketing sources are doing.

There are ten top mobile-advertising platforms that you'll want to pay attention to – Google's AdMob, Millennial Media, iAd (from Apple), Flurry (in addition to a free analytics solution, it also has a big ad network), inMobi, Chartboost, MoPub, Amobee, HasOffers and Euclid Analytics.[6]

In the next chapter we'll go into a lot more detail about advertising and investigate many more channels, and we'll look at how testing different campaigns, messages and calls to action will make your user-acquisition strategy more effective and efficient.

Driving Downloads

So, given that you're unlikely to have much budget in place to spend on advertising (but at least you know how it all works and the basics of how to track everything), what are the other options you have to get those precious downloads started?

Let's start off with the mechanical things you can do: publicity and social media.

Publicity

PRESS KIT. You should already have a press-kit section on your website (we talked about this earlier). Make sure it is easy to find and includes your logo, screenshots of your app in high definition, a clear company story and profiles and photos of the founders and key people in the team.

PRESS RELEASES. Make sure you do your homework and check out some examples used effectively by other apps. Once you feel that you actually have something newsworthy, get your press release out on the wire – and, if you've actually put together something useful, interesting and accurate, it will get picked up. It usually costs only about $200 to do this yourself with PRWeb, or up to about $1,000 if you use some of the other targeted wire services such as Business Wire.[7]

DO THE LEGWORK. Reach out to journalists yourself. Just like the rest of us, they are lazy. If you present them with a good story, and a personalised email outlining what you're doing, there is a good chance you'll be picked up. In the process, pull together a list of blogs in your relevant sector (technology and app news) as well as popular blogs that love reviewing apps (there's a myriad of those), and then spend a few days writing a lot of emails – and be pre-pared to reply to a lot (and also get ready for a lot of interviews – it's increasingly common to do them over Skype and Google Hangouts). So get out there!

Social media

You should already have all your social-media accounts set up. With the aid of a tool such as HootSuite you can also manage – and publish to – all those accounts from the one place (definitely a great time saver). So get ready to put some good content out there. The best thing about tools such as HootSuite is that you can com-pose the content in advance – e.g. five posts over the weekend – and then schedule them to be published 'automagically' during the week.

One good piece of advice about social media: you want to get into the habit of paying attention regularly and often in the early days. Pick a day of the week, and then systematically follow 100 people that are pertinent to your business. With Facebook, follow all your real friends and invite everyone to follow your page. Social media isn't for the shy! If you make this systematic you'll get into the thousands of followers in no time. One thing that I try to do is make it the last task I do before I go to sleep (I do it from bed). Five to ten minutes a day definitely adds up.

So these are the basics – and you should feel great now that you have all this in place. As we progress it's all about how to build this into a growing, expanding machine, with passionate people responsible for every aspect.

Chapter 11

Is Your App Ready for Investment?

You have caught the entrepreneurial bug. You want to make your app work and take it to the next level. You have your founding team in place. You've been testing your prototype app with real users. You have some encouraging data. This is the point where you want to consider outside investment. Why? Investment will help you grow faster; it will enable you to hire more people, spend more on advertising and get more people using your app.

But what does your app look like right now? And how much data is enough to persuade investors to take a bet on you? Unfortunately, there are no set-in-stone rules at this point, because different teams chasing different types of business models are going to be able to deliver some wildly different results.

Let's have a look at some examples, and see what some apps have been able to achieve.

At this seed stage Evan Spiegel and Bobby Murphy, the founders of Snapchat, were still living at home. They were a team of two. They were straight out of college with zero experience. They hacked together their app and submitted it to Apple's App Store. At best it was average-looking; someone harsher might have called it ugly. It was certainly not very complex. But it did have a basic user registration function, allowed you to add friends, and also allowed you to send messages to those friends that disappeared after 10 seconds. The app was very much an MVP, a minimum viable product. They invited a bunch of college and high school students to use it, and they invited their friends. Their user-acquisition metric was self-sustaining (because of the inherent network effect of users inviting their friends). Their user activation (effectively creating an account) converted at close to 100 per cent because it was a super-simple registration (just a user name and password). Their retention started growing as users began messaging one another and sending snaps (or photos), and their referral metric was off the charts because it was baked into the app (the app works better when you invite more friends to it). It was basically a perfect storm.

At this point they *had* to raise money because they could no longer pay the server bills to support all the traffic the app was generating.

Let's look at a more realistic scenario. You've built a pretty solid prototype app. It's well designed, contains all the requisite features and is clear and simple to use. You should have been able to attract a few hundred friends and family to test it out, and ideally a few hundred more by contacting blogs and various email lists, and even have invested some money in paid advertising and downloads.

If you're looking for investment, you should be able to show a solid download-to-user-acquisition rate of around 80–90 per cent (always shoot higher). Depending on whether you've targeted your users well – and whether the first-time-user experience is compelling – you should be able to activate upwards of 50 per cent of your users. (In the very early days of Hailo this was around 25 per cent until we introduced a first-time-user 'tour' that explained how the app worked. The tour shot our activation rate above 50 per cent.)

In terms of retention, this depends on whether you're a gaming app (where the leading apps keep people coming back daily) or an e-commerce app (typically, users come back 1–4 times per month). It's tricky giving general metrics here – so you're going to have to research apps in your specific market. In terms of referral, unless you're a communications-type app, where every user should be inviting lots of people if your proposition is compelling, you're generally going to see only a small proportion – generally 2–5 per cent – of users who actually invite a significant number of others to download your app.

At this stage, any metrics around revenue are going to be great news – and, frankly, a bonus. The expectation won't be there from angel investors, so, if you can impress them with actual dollars coming through the door, this is only going to help you.

If you want to hold yourself up to the highest possible bar – like WhatsApp – then you won't need funding at all. Jan Koum and his cofounder started WhatsApp with their own money – they are both programmers – so built the initial app themselves. And, due to a combination of a great app and the fact they charged $0.99 to users to download it, they were generating revenue from their first day in the App Store. Their proposition of unlimited messaging for a one-off charge struck a chord and propelled their initial user adoption. They avoided the hassle and annoyance of raising any financing early on. And that is by far the best – and most profitable – option.

If you aren't able to create your own perfect storm, then you will have to go and find some money. That money will give you the ability to focus full time on developing your app – and getting the leanest possible team on board to help. You'll need enough money to get to the next level, which involves improving your app to the point where you have achieved product–market fit and have sufficient users to test your ability to generate revenue.

So let's have a look at the best place to start looking for some early funding.

Angel Investors

Angel investors are the people – quite often friends and family – who provide that initial injection of cash to get the first version of your app off the ground. It's their cash that helps you, the founder, stay off the street and keep the dream alive. Just like any other investor, angels take a percentage ownership of your company in exchange for cash. Their goal is to see your company grow in value, and therefore have the value of their ownership stake increase significantly.

Angel investors really invest in people, as, at this stage, it's very much about feeding and enabling a vision. At such an early stage there is little else to back apart from the founder, their track record and the quality of their idea. Being an angel investor is not really about making an extraordinary profit. In all honesty, it's horribly risky, since most startups and entrepreneurs fail. So in many senses it's charity.

That said, there will be stages in your career where you're a charity case, and where a small amount of cash and belief from an angel means the difference between you remaining an entrepreneur and going back to slave as an analyst or consultant. So it's a good thing.

But where do you start hunting for angel investors beyond your friends and family? The aptly named AngelList is a great first port of call. Funnily enough, it's a truly fantastic startup itself (started in none other than Silicon Valley). In September 2013 the company landed a hefty $24 million investment to keep helping the startup ecosystem. It is effectively a social-network/dating site for entrepreneurs and investors, mainly focused on the early rounds.

Get details about yourself – and your company – up there and start meeting people who are interested in investing anywhere from $5,000 to $100,000 in your fledgling app. It's by far the most impressive site of its kind, and it's becoming increasingly global.

Angel groups are very localised, so you'll need to do some research in your local capital city or tech hub to see who is active. Check out mybilliondollarapp.com as well – I've put together a few great collections in 15 major cities to help kick-start your search.

Venture Capitalists

You've probably come across the name before, but what are venture capitalists (or VCs)? VCs are simply very experienced and knowledgeable (well, some of them) bankers. Just like normal bankers, they will lend you money. Unlike normal bankers they *don't* charge interest! Amazing, I hear you say – but instead they can take a direct ownership share in your company. They give you cash, you give them shares. That makes them more like spouses – you are pretty well married for the duration of your business – so you'd better like each other.

Many people think that VCs are merely opportunistic investors who add little real value to your company. While that sentiment may be an accurate assessment of many average VCs, the best VCs do provide excellent advice and value. They are vultures in the sense that they pounce on the best opportunities, but they also have the power to help you grow and develop your business faster – and better – than would otherwise be possible.

Conventional bankers make money by lending you some, and then charging you interest on the amount that is outstanding. Often, they make sure that they have some collateral that will be given to them in case you should not be able to repay the cash (like your car, house or children). These bankers are pretty risk-averse, and most people keep paying their loan or mortgage payments every month. Since the risk isn't that massive, the returns are equally conservative. These normal bankers make money by lending small amounts, but to millions of people. They profit from scale.

VCs, on the other hand, love risk. Their business is in making big returns on big investments. If a VC invests $100,000 in your business, they don't want to see $105,000 after one year, or $121,000 after four years: they want to see a multiple – times 2 or even times 10 if you've really knocked it out of the park.

For every 10 companies that a VC invests in, their goal is to have one company make a tenfold return (i.e. for an investment of $1 million, a return of $10 million), three companies to return five times the investment, three or four to break even (i.e. the VC recovers the initial investment that they put in) and then, on two or three companies, they expect to get no return (effectively lose their money). It's a pretty scary model when you think about it.

At this stage, this a very general picture, and naturally hugely simplified, but I just want you to understand the ecosystem. And I want you to understand that VCs are a tool that you can use to help achieve the goals you have for your company.

It is a lot easier to accelerate towards your billion-dollar vision when you have the money and resources to grow at a superfast rate, rather than having to spend 20 years slowly building a company. But, naturally, everything comes with a price.

At this seed stage you probably won't be ready to start talking to VCs and you will mainly be dealing with angels. But, if you're experienced and have a history of building up companies and apps, there is a class of 'seed-stage VCs' who will be very interested in talking.

Sharks as Friends

To describe VCs as 'bankers with an overzealous appetite for risk' is perhaps an oversimplification. I am lucky enough to have forged many strong friendships in the venture community, and I have to say they're pretty damn smart people.

Even if you're not yet looking for their investment, make friends with them. They see the insides of numerous successful – and failed – companies that you never will (unless you become a VC yourself). The more experienced the VC, the more insight they can give you – and, if you become friends with them, you get all this information for free.

So what can VCs help you with at this stage? First, find one who knows your market inside out. If, say, your app is a game, a VC who has previously invested in a gaming company will know the ins and outs of the sector. They will know the leaders, their cofounders, their business models, their competitors. Since all startup companies are, by the very nature of their youth, private, it is very hard to get any information about them (except the information they *want* you to read via PR – and we all know that we should be quite careful about how much we believe that). Their job is to get data that is hard to find publicly.

Helpfully, the venture-capital community is pretty small, and the very best investors – such as the Accels, KPCBs, Sequoias and USVs – all tend to see a lot of the same companies pitching for funding and hence tend to see a lot of confidential information, and thus they have a lot of privileged insight. So, if you know people working at these investment companies, they can give you a very good steer – and brilliant advice – while at the same time not divulging any privileged information.

Accelerate or Incubate

If you're looking for another option, then an incubator or accelerator might be for you. A startup incubator is basically a fixed-duration programme to get a basic startup idea off the ground, with support from mentors and typically with office space for the duration. Accelerators typically take over at the next phase, taking an idea that is pretty well developed and giving it the extra support: money, mentors and office space to get to the next stage (typically external funding).

They seem to have blossomed all over the world, but, despite their popularity, there are still only a handful that yield any kind of meaningful results.

These programmes focus on providing great entrepreneurs with the opportunity to refine their ideas, their business models and their products before putting them in front of a set of groomed investors.

In a very condensed timespan – typically not much more than three months – you meet with hyper-successful mentors daily and weekly, and are expected to work day and night. You can make magic happen in that kind of environment. It's a great alternative to just going and chasing seed investors via AngelList (or good old-fashioned hitting the pavement).

I have a few friends who have graduated from two accelerators – Y Combinator and Techstars – and the feedback has been mixed. Americans love having seals of approval from prestigious institutions, so the programme seems to yield great results for them, but the Europeans seem a bit more tepid about the format. Personally, I think you get back what you put in, and having access to their network of alumni and their inside knowledge is a precious resource.

Let's have a look the three top programmes.

Y Combinator is probably *the* best seed accelerator. It was started in March 2005 and has funded more than 500 companies – including Dropbox, Airbnb and Stripe. It provides seed money, advice and connections during two three-month programmes each year. In exchange, it takes an average of about 6 per cent of the company's equity. It receives about 1,000 applications for each class – and accepts about 38 – so it has a 4 per cent acceptance rate.[1]

Techstars is another US outfit that is jockeying for that number-one startup accelerator crown. It receives thousands of applications per year, but invests its money and time in only about 10 companies per location. The programme has, at the time of writing, seven locations: Boston, Boulder, Chicago, New York City, Seattle, London and Austin. It invests $118,000 in each company it funds through $18,000 in seed funding and an optional $100,000 convertible debt note. Techstars is backed by 75 different venture-capital firms and angel investors. They provide three months of intensive top-notch mentorship and the chance to pitch to angel investors and venture capitalists at the end of the programme. The companies average over $1.4 million in outside venture capital raised after leaving Techstars.

Seedcamp was the leading accelerator in Europe, but Techstars ended that when it landed in London in 2013 (it merged with Springboard). Seedcamp claims to be the most connected international seed investor in the world and Europe's leading micro-seed investment and mentoring programme. Since launching in 2007, it has supported more than 90 startups in Europe. It invests in 20 startups globally annually.

I've included a lot more information about other programmes on mybilliondollarapp.com, including how to break into Dave McClure's 500 Startups and the Entrepreneur First programme, a London-based accelerator that attracts applications from one in three eligible Cambridge Computer Scientists.[2]

Chapter 12

How Much is Your App Worth and How Much Money Should You Raise?

'Cash flow is more important than your mother'

Before approaching any investors, you need to figure out how much your app is worth. This will feed into your decision about how much money you're trying to raise. To reasonable, everyday, normal people, the business of valuing a tech or app company is going to make little or no sense. Even to seasoned business people without much exposure to the tech world, it causes quite a few raised eyebrows.

As we saw earlier, the more savvy technology investors want to buy a share of your company. What they really want to own is a *certain percentage* of your company – and everyone makes money when the valuation of the company increases. Coming up with the very first valuation is the trickiest – in some senses – because it is by definition a stage where little has so far been delivered.

What ends up happening tends to be a little backwards. Let's walk through a couple of golden rules.

At this early stage, investors want to own anywhere between 10 and 25 per cent of your company in return for their investment. They want to make sure they are compensated for the risk they are taking by giving you money so early in the process, and they want to protect against their investment being diluted by new investors coming in during subsequent funding rounds.

Investors want to make sure that you have enough cash to get to the next stage – a well-flushed-out product with great market fit. Some teams require only a small amount of money to get to that point, and some apps require a lot more, because the products are more complex to get off the ground.

So what ends up happening is a negotiation. If you are a small team with only a bit of experience and need 12 months of runway to get your app out there to market, a $200,000 investment might get you there. And your valuation will probably be at the lower end of the spectrum, around $700,000 or so.

If your team is more experienced and has a track record – and there are a few more of you – and you need a bit more runway to prove your product (Hailo needed to build both the driver and passenger apps very quickly), then you might need to raise $1,000,000 or more to get you there, and perhaps you could agree a valuation of $4–5 million.

According to the Angel Capital Association, the median pre-money valuation (the valuation of your company before an investor puts in their money) of companies that are not yet generating revenue was $2.75 million in 2012. This was an increase over 2011, when it was $2.1 million, and an even larger increase over 2010, when it was $1.7 million.[1]

This means the valuation of a typical Silicon Valley startup – at the angel or seed stage – is $2.75 million. On average, these companies raise about $750,000.[2] That means that, after that cash is added to their bank account, they are worth $3.5 million on average, and the investors own 21.42 per cent. So, while raising money this way costs you quite a bit of ownership of your company, there really isn't another way to raise that amount to grow your company so quickly.

Convertible Loans

It is becoming increasingly popular for startup companies to raise initial capital by way of something called a 'convertible loan'. A convertible loan is first and foremost a loan: it involves borrowing money that has to be paid back with interest. The difference is that the conversion feature gives the investor the option to convert all (or a portion of the outstanding amount of the loan) into equity in your company.

Why has this approach become more common? Basically, it avoids the valuation discussion and potential to 'get it wrong' (there can be negative consequences to having a very high valuation, as it can limit your choice of investors). In exchange for the flexibility offered by a convertible loan, investors will end up getting a discount when they convert the loan amount into equity in the next funding round (typically 20 per cent) or ask for a valuation cap (which can vary). Although the idea of a valuation cap seems to be penalising, the final valuation you receive will definitely be higher than if you sold equity earlier on via the non-convertible-loan route. So in a number of ways, it is a win-win. Many investors, however, would rather have the equity in their hands.

Annoying Legally Things

After a lot of legwork, you've convinced your friends and family and a couple of angel investors that your great idea is worth investment. And maybe you've even convinced a venture capitalist who specialises in seed-stage deals. Now things are serious – and here's what you need to do next.

First, you need to incorporate a company. These days that's easy. Head over to a site like companiesmadesimple.com (I've used these guys with great success). You can incorporate your very own company with a credit card, five minutes of your time and £30. It's all done online – and you get a digital certificate of incorporation, a company number and a list of the initial shareholders of the company. Keep the structure simple when you do this initial incorporation since it will change when you complete this first funding round. There's lots of fun with the lawyers coming up.[3]

So you're the proud cofounder of a fresh new corporate entity. Congratulations! The next step is to go and get a bank account – relatively easy these days. Now, get into the habit of using accounting software (Xero – pronounced 'zero' – is super-functional, and also beautiful). You'll also need an accountant on hand to make sure that all your annual accounting paperwork is completed and filed with the tax people. Having a tight grip on your finances is key, because cash flow will be the most important thing you manage over the coming years. If you keep all your transactions – purchases, invoices and receipts – up-to-date in this system it will alleviate a lot of headaches. I still remember what John Doerr, the famed venture investor who was an early backer of numerous iconic Internet companies including Google, Amazon, Flipboard, Zynga and Netscape (and has a personal fortune of more than $2 billion[4]), said back in my senior year of college: 'Cash flow is more important than your mother.'

Pretty soon you'll need to line up a proper accountant to make sure your books are in order, that you're paying whatever taxes and charges you need to pay and that you're taking advantage of all the tax breaks you can as a fledgling company. They will also be helpful throughout the fundraising process. In case you need to find an accountant, I've included a bunch of useful resources on mybilliondollarapp.com.

Get Protection

From the get go it's a good idea to make sure that you have official agreements in place with all the people and companies you're working with. This means signed contracts, but it doesn't necessarily mean a high legal cost.

First, make sure that when you start having serious conversations with other parties (contractors, freelancers, potential employees or agencies) you have them sign an NDA (Non Disclosure Agreement). Requesting this makes you look a bit more professional and also gives you a small amount of protection should someone want to steal your idea. If someone doesn't want to sign it, then that could be an indication they have an ulterior motive. Do note, however, that VCs will never sign NDAs so don't bother asking (it would basically preclude them from talking to entrepreneurs with similar ideas – which is what they do on a day-to-day basis).

Second, you need to ensure that you clearly own all your intellectual property, in other words, you need to make it clear that all the work you contract other parties to do belongs to your company. For example, if people are designing logos, illustrations or videos for you, you need to make sure that work belongs entirely to you. If people are developing an app, website or other piece of software, then you need to make sure the source code belongs to you. On a practical level, this means you need to have signed agreements with anybody who does any work whatsoever for your company.

With a clearly articulated and signed agreement in place there is no room for doubt. This will prevent unscrupulous, or just opportunistic, people from trying to take advantage of you. I have encountered rather nasty situations whereby people working from a verbal contract have refused to deliver a design or a piece of code because the price or deadlines were disputed. I have also seen situations where contractors refused to hand over source codes once a piece of work was completed – again because there was no clearly written contract.

Do yourself a favour and get contracts drawn up. I have included a number of legally approved contracts in the legal section of mybilliondollarapp.com for you to look at and use.

Founder Vesting

One of the critical things you're going to need to understand as you go forward is the concept of 'founder vesting'. This may or may not come up if you're working only with angel investors at this stage. These guys are typically a bit more calm. If you're dealing with a professional investor or VC, this definitely *will* come up.

The concept revolves around the fact that professional investors want assurances that you and your cofounders are going to hang around in the future. A founder-vesting schedule is basically a timeline of your earning full ownership of your shares in the company. Typically this means you will need to stay and work with your own company for four years (a normal period in the United States, in Europe it tends to be three years) for all your founder shares in the company to vest (i.e. effectively become your property), at which point you own them outright.

On the surface, this sounds a bit nuts: you work your guts out for your company and an investor wants to tell you that you have to earn it back! But, it's not quite as nuts as it sounds. Let's have a look at the simplest example, and why it looks after your interests.

Let's say you and a friend cofound a business. You each own 50 per cent. Six months in, your friend decides that he wants out of the company. But you have already built an app, contracts have been signed and you've personally invested a lot of money in the company (while the friend has invested nothing but his time). Legally, you both own 50 per cent. If he walks away, he still owns 50 per cent. You get stuck running the company, but you, too, own only 50 per cent. That sucks. And there's not a lot you can do about it.

If, on the other hand, there was an agreement in place whereby both of you had founder vesting over a period of four years, and for every month you stayed with the company more of your shares vested, then we have a different situation. Say your cofounder again upped and left after that same six-month period. Instead of pocketing 50 per cent of the entire company right then and there, he'd pocket only six months' worth of the four years – or 12.5 per cent of the 50 per cent that he would have owned (since he stayed only one-eighth of the vesting term). So he could walk away with 6.25 per cent – which is what he earned by staying that long. You may not be in love with this situation either, but it's a lot fairer than losing control of an entire 50 per cent.

So, while professional investors will try to get you to think in this way, you may want to consider this option even before investors come to the table. It's a good way to protect yourself – especially if you have limited experience with a new cofounder.

The only exception, you may argue, is if you are the sole founder. Given that you have everything at stake – especially if you've personally invested a lot in the venture – you have a chance to remove this provision.

Signing a Deal

Provided you have concretely agreed a valuation with all your investors, at this stage everything comes together in a neat legal document: how much your company is worth, how many shares there are, who owns how many, and which investors are buying how many at what price and with what conditions attached.

If you thought getting a mortgage was tough, it's nothing compared with having investors. It will change your life. But don't worry too much about that at this point – the investment and entrepreneur community has come to a consensus on what a good and fair deal looks like for both sides of the table, and, as a result, seed-investment-stage contracts are relatively standardised.

Check out the following resources:

- **SERIES SEED DOCUMENTS.** You can find them at seriesseed.com (download the documents for free on the website). These documents have been prepared by Fenwick and West – a top Silicon Valley law firm. The documents are very widely used – so do make yourself very familiar with them. Why is this important? If an investor says that the Series Seed documentation is bogus or nonstandard, then you know you have an issue with that investor.
- **TECHSTARS.** We've talked about this outstanding accelerator/incubator before. It also has its own standard set of seed-investment documents.[5] You can check them out on the Billion-Dollar App website. It's worth reading through these documents and seeing how/where they differ from the Series Seed documentation.
- **THE FOUNDER INSTITUTE.** FI is probably the world's largest entrepreneur training and startup launch programme – with programmes in more than 50 cities. It has a version called the New Plain Preferred Term Sheet.[6]

If you want to know how all these documents compare, then first ask that lawyer friend, or visit the legal section on mybilliondollarapp.com, where I provide a lot of great resources and documents.

If you're serious about running an app startup, you will want to make great friends with lawyers. They are nice people too – and can be a great resource to call on when you're lost and not sure if you're getting screwed. It's great to feel that you can ask a question – and you won't get billed $500 for the privilege.

The Core Documents

In this first funding round, you're going to feel quite overwhelmed. The good news is that you're unlikely to run into any major problems if you follow the basic rules I've set out in this section. The documents and conditions are pretty standard – so all you need to worry about is getting a great valuation.

Everything starts with the *term sheet*. This is basically a nutshell agreement – no more than a couple of pages – that summarises how much an investor is willing to invest and under what conditions.

If you sign this agreement, then you're committed to collaborating with your investor to work through the details of negotiating a longer version of that agreement.

As part of the actual agreement, you'll have four main things to agree:

- **Restated certification** or **articles of incorporation**, which are typically needed to properly document the existence and treatment surrounding the new class of preferred stock that investors insist on;
- **Preferred stock investment agreement**, which documents the actual details of the investment;
- Bylaws that describe in detail how the **company will be governed** – read this very carefully; and
- The **board-member election consent**, which elects the board member representing the investors.

So there aren't a huge number of documents to get your head around. All are available online, all are pretty standard. If you wish to do a bit of further reading, I thoroughly recommend *Venture Deals: Be Smarter Than Your Lawyer and Venture Capitalist* by Brad Feld and Jason Mendelson. It's a very helpful book.

Here's some good advice:

- You should always negotiate a term sheet directly with your investors and not hide behind a lawyer. If you cannot get along with your investor at this early stage, then it is best to find that out now, rather than later.
- Read and understand everything in the term sheet. But, when it comes to the closing docs (which can be very long), sit down with your lawyer and ask them to explain them in their entirety, focusing on the clauses that could be written more favourably to the startup.
- Get a good lawyer. Ask around for lawyers who have worked on both the entrepreneur's and investor's sides. This way you get a much richer perspective. Don't assume your lawyer is good just because he works at a big law firm. I've dealt with fantastic lawyers at smaller firms. Dig into their personal experience and see which VCs they've worked with before and which other startups they've represented.
- You probably won't be able to tell the difference between good legal advice and bad legal advice at this stage. So ask lots of questions, pester people, and don't sign anything you're not comfortable with. And remember: this has been done many times before, and there's a trove of information about seed deals on the Internet.

Hopefully all this makes the process seem a bit less daunting. The one final part that will be painful will be the cost. Be prepared to part with anything from £10,000 to £20,000 for a simple seed-round investment. And, yes, it is standard to pay a certain amount towards the investors' legal bills as well – ludicrous but true.

Summary

The big – and simple – goal we set at the beginning of the chapter was validation. Were you able to build up a core team, which might just be you and a cofounder? Did you successfully translate your killer idea into a design, wireframe, prototype and product? And is it beginning to wow your early users?

And the toughest part (on a nonexistent budget): were you able to get your app out there and start demonstrating some traction and, most importantly, gathering feedback from early users to help you to start refining your app? Unfortunately, all these steps are not linear; nothing is ever as simple or structured as you would like. The app prototype might come before the cofounder, and users may be clamouring to download your app before it's completely polished.

In the end, getting to a point where you're talking with investors is a great place to be. If you're in a position where someone wants to give you money to take your app to the next level, well, that is both inspiring and petrifying.

Once you reach this point, you've taken the first step on the ladder to creating a real business. Next, we focus on using the money you just raised to get to the point of real product–market fit, focusing relentlessly on the wow factor and then making sure you grow both your user base and a business off the back of that achievement.

STEP 2

The Ten-Million-Dollar App

Achieving Product–Market Fit and Raising Series A Funding

Going from a Million-Dollar App to a Ten-Million-Dollar App

App

- Your prototype app is impressing people and they're using it. You have the beginnings of the 'wow' factor.

- If your business model is simple (gaming, Software as a Service), then your Version 1.0 app is out there in the app stores.

- If your model is more complicated (marketplace), then you have a solid proof of concept, and will use your investment to develop it further.

- Throughout this stage you're going to need to be 100 per cent focused. The only thing you care about now is achieving strong product–market fit – ensuring users love your app so much they are willing to pay for it.

Team

- The founders are in place and the core team members you earmarked can now join full time because you have some funding.

- You're still a pretty small team, but all core functions are covered. As the product improves and traction picks up, you need to think about the next big hire you'll probably be making to drive more users and revenue: the VP of marketing.

Users

- From a small user base in the hundreds, you'll start to grow quickly into the thousands of users.

- You'll be developing a solid and reliable user-acquisition strategy. You'll figure out what the cost of acquiring a user currently is (and what needs to be done to make your business profitable).

- You'll also start to put in place the basic tactics around retaining users and keeping them coming back and using your app.

Business model

- What was a basic business model will now evolve quickly. You need to make sure that it works in practice as well as in principle.

- You'll validate it by constantly testing with users, measuring and analysing their behaviour, and experimenting with new features to get them using – and paying – more.

- The quicker you achieve product–market fit, the quicker you can start testing revenue strategies. But stay lean and mean, because getting to product–market fit can take a lot longer than expected.

Valuation

- Demonstrating that you can build an app that users love to use and love to pay for is hard. It requires patience, experimentation and perseverance. But, once you deliver the goods, investors will reward you with on average a $10 million valuation.

Investment

- At this point things get serious. You're going to move into the world of professional venture-capital investors. That means long agreements and lots of legal paperwork.

- But, if you really want to ramp your business, develop your app, build your team and spend on marketing, then this is the quickest – and potentially smartest – way to raise around $2–3 million.

Chapter 13

HMS *President*

The cab driver didn't really know what to think. He'd been using this new-fangled Hailo app on his iPhone for a few weeks. It was proving to be pretty useful: he'd used it to find new fares, used the analytics feature to discover that on Tuesdays he was empty 54 per cent of the time and even accepted his first credit-card payment. What was not to like?

This morning, however, the app started buzzing and clanging. It hadn't done this before. He examined the screen more closely – it was a customer hailing him via the app. Someone wanted to be picked up. *Oh shit*, he thought. He had no idea what else to do but press the green 'Accept Job' button that took up most of the smart-phone's screen.

Immediately the screen changed. It now showed an address: HMS *President*, Victoria Embankment, London. There was no street number. But there was a map. And so off he went. Within about five minutes he was at the location marked on the map, and sure enough there was a huge boat – the HMS *President* – moored right there, smack in the middle of the Thames. Alas, there was not a person in sight. Was this Hailo app working correctly?

It was right then that I felt my phone ring. I'd been watching my iPhone's screen – with the Hailo app open – tracking the progress of the taxi I had hailed. This was the very first Hailo job where the driver hadn't been warned ahead of time. I could see he was outside and now he was calling me, since we'd designed the app to prompt the driver to get in touch with the passenger when they arrived and no one was waiting outside.

'Hello, it's your Hailo driver Phil,' he said.

'Fantastic, I'll be outside in a minute. You're in the right place,' I replied.

And so began the long process of testing our app on real users. While we had done everything we could to ensure we designed and built the right app, nothing except testing in the wild would validate that. This was the beginning of the process to validate whether we had achieved product–market fit with our app.

There was a reason the cab driver was called to such an unusual location that day: a ship called HMS *President*, a World War One destroyer that went into service in 1917, would serve as the very first office for Hailo. (After helping the war effort, it was turned into a permanent attraction on the River Thames – and operates as an office and events venue.)

It was from here that we conducted months of testing with both drivers and passengers. The atypical location was a good test for drivers to find a bizarre passenger pickup location and also served as a test of the accuracy and usability of the app for passengers.

Throughout the testing we made lots of changes, worked with countless drivers and passengers and really tuned our service – and app – to the point where people loved it. This in turn gave us the foundation to raise a strong Series A funding round – one that would really help us grow the service through London, and beyond.

*

To get to the million-dollar stage, you built a functioning prototype version of your app, and a decent number of people liked it. You've shown a lot of promise, which is why you've been able to land your first round of investment. You've cut your teeth hacking out a deal with investors, and you've quickly learned how easy it is to spend too much on lawyers. But now – with the contracts signed – the cash is in the bank. And that means the real fun can start.

Take a moment to freak out about what's coming – and then relax. It's pretty impressive to make it to this point.

The money in the bank now serves one core purpose: making sure you get product–market fit. There is nothing more important at this stage. Once you have tuned your app so that it resonates with what your target group of users wants and needs, you will have the foundation, the perfect chemical formula, to ignite a billion-dollar business.

Even if you were able to hit some of the truly spectacular metrics and get traction early on, you still need to make sure that this *initial* demand can become a real business.

You begin this chapter as an incorporated company, just a handful of people. It's been a labour of love, an exercise in calling on every favour you can. Now, with some funding, you can finally pay those lingering invoices – and hopefully persuade a few key people to join you full time – and pursue the goal of delivering your wow moment.

It may seem unbelievable, but at the end of this chapter, if everything goes to plan, you will have an operational app that will be delighting users and gaining real traction out there in the real world; you'll be making real revenue (ideally) and your team will be growing, probably to around 7–10 people (though this depends on how much revenue you're earning or how much money you've raised). If you hit the product–market fit faster, then you could be a lot bigger than that.

Whatever happens, it's going to be a big transformation.

Chapter 14

Make Something People Love

'The life of any startup can be divided into two parts – before product–market fit and after product–market fit'

#BILLIONDOLLARAPP

Marc Andreessen was the co-creator of Mosaic, the first widely used Web browser; he was cofounder of Netscape; he is the cofounder and general partner of Silicon Valley venture capital firm Andreessen Horowitz. He sits on the board of directors of Facebook, eBay and HP, among others. He's a pretty smart guy who knows a thing or two about tech.

Andreessen explains on his blog,[1] '... the life of any startup can be divided into two parts – before product–market fit (BPMF) and after product–market fit (APMF).'

He continues,

> When you are BPMF, focus obsessively on getting to product–market fit. Do whatever is required to get to product–market fit. Including changing out people, rewriting your product, moving into a different market, telling customers no when you don't want to, telling customers yes when you don't want to ... – whatever is required.

Product–market fit means first, being in a good market and, second, building a product that can satisfy what people in that market want. Without that you're not going to experience explosive growth.

We've talked in depth about developing a billion-dollar idea – a large part of that is being in a large, growing market. By focusing on the app world, you've already solved half that equation, you're in a huge, growing and international market, and now you just need to build something that people want.

The core thing to remember about your app is that it should be something people want, which means getting to the very bottom of the problem it solves and how elegantly, quickly and painlessly it solves that problem.

You can definitely feel when you haven't quite reached the point of perfect product–market fit. Your users don't seem to get the value out of your app, there doesn't seem to be much word-of-mouth growth, usage isn't growing that fast, and you're receiving rather flat publicity (no one is getting to the point of wow). You're still stuck in the limbo world of the 'me-too product'.

The good news is that you can definitely feel when you *have* achieved product–market fit: users are downloading your app in droves, you can't spin up servers fast enough to support demand, and iTunes and Google Play are depositing more and more money into your bank account. You find it hard to hire support staff any faster, and you're tired of talking to reporters on a daily basis. And then you win Apple's App of the Year Award. Success becomes pretty obvious.

But this doesn't happen suddenly – and it certainly doesn't happen overnight. There is a path from the good, solid app you created to get to million-dollar stage – a minimum viable product out there in the hands of users – to one that users *love*.

Let's make it a bit less ephemeral. I am an engineer, after all. I'd like to be able to measure something that tells me whether I've reached product–market fit – or at least whether I am approaching it – via an objective measure. I like the way that Sean Ellis – founder of Growth Hackers – describes it:

> I've tried to make the concept less abstract by offering a specific metric for determining product–market fit. I ask existing users of a product how they would feel if they could no longer use the product. In my experience, achieving product–market fit requires *at least 40 per cent* of users saying they would be 'very disappointed' *without* your product.

He continues, 'Those that struggle for traction are always under 40 per cent, while most that gain strong traction exceed 40 per cent.'[2]

To me, that statement feels right. In essence you need a major part of your user base to love what you're doing. When we took Hailo to market at the end of 2011, I spent an inordinate amount of time surveying users (and giving out lots of in-app promo codes for £5 off your first ride to friends, family, random people on the street and anyone who would listen). After hundreds of feedback sessions, Hailo was getting a solid 75 per cent of people loving the service. In fact people were quite incredulous that an app was able to deliver such 'magical' results.

The Aha! Moment

The first time I used the Hailo app to hail a taxi was an emotional moment (really, it was). But the real wow would come a few weeks later when we started testing it with our families and friends. Even though it wasn't 'truly' out in the wild, a big group of people were sending ecstatic feedback to us. We could see we'd hit product–market fit in London, and now we just had to focus on making the system robust enough to deal with huge volumes of users in the real world.

The first time I experienced the Square register app I was truly wowed. It was in a small café in New York. Having just arrived from London earlier that morning, I didn't have any cash, so I asked whether they accepted cards. The girl behind the counter happily replied yes. When she swiped my card through the white shiny dongle attached to an iPhone I couldn't help but ask her what that was. 'It's a new app called Square. It turns your iPhone or iPad into a register,' she said. 'And you get this white swipey thing for free.' She then asked me to sign the screen of her iPhone with my finger. I thought that was rather slick (though not particularly secure).

'What's your email?' she continued. 'I can email you the receipt so we don't have to kill any trees unnecessarily.' That pretty much closed the deal for me – this app was amazing. The potential of something so simple and elegant made me immediately think about the impact it would have as it reached other cities around the world.

This is the bar that your app has to hit to become a billion-dollar success. It's a high bar – but it's not an unreachable one. There is a process that you can adopt to get to wow, and I'm going to dissect the best ways to do that throughout this section.

Engines of Success

So, while your principal focus is building an app people love, there are a number of other things you'll need to be working on simultaneously.

Once you've achieved product–market fit – part of which is validating your fundamental business model – your focus will quickly evolve to rapidly growing your users and tuning your business model. That means that you'll need to start putting in place the two engines that will power your business:

1. GROWTH ENGINE. You need a systematic way to grow your user base. You need to reliably generate downloads, convert those downloads into valuable, revenue-generating users and then keep that cycle going. That involves being able to test new user-acquisition channels, test their effectiveness and then figure out what kind of lifetime value (LTV) they generate. If you can figure out how to repeatedly deliver a customer acquisition cost (CAC) for a user that is below the LTV, you have a profitable, sustainable business.

2. REVENUE ENGINE. Similarly, you need to validate how your app is going to make revenue. Is it going to be a consumer-audience app that will monetise via advertising? Will it be an e-commerce or marketplace app that generates revenues via transactions? Will it be a SaaS or enterprise app generating subscription revenues? Or will it be a gaming or virtual-goods app, generating revenues via premium downloads or in-app purchases? Part of achieving product–market fit is demonstrating – with sufficient data – that users are super-happy to pay to use your service.

Throughout this stage of the journey we'll focus on how best to use your seed financing to achieve these goals and set the foundations for a solid app company – and put you in a great place to seduce some serious professional investors to allow you to expand your team and accelerate growth.

How to Deliver Wow

Paul Graham – one of the founders of Y Combinator, Silicon Valley's top startup incubator – offers a great morsel of advice after decades of experience delivering software: 'Don't build something clever, build what people want.'

The nature of technology – and software companies in particular – has evolved quickly over the last decade. Open-source software – along with easily accessible services such as payment, mapping, messaging (and many others) in the form of APIs (application programming interfaces) – allows any developer to build powerful programs.

As a result, anyone can now build something *clever*. It is harder to focus on building what people *want*. Software companies, therefore, have shifted from big teams focused on developing the basic building blocks to small lean teams focused on delivering solutions – products – that solve an increasing array of everyday problems. This move to creating solutions rather than software has forced the people who actually make software to better understand people and their behaviours.

Product-Centricity

The term that describes this shift is 'product-centricity'. To be product-centric is to be maniacal about building the best *product* out there for your users. Users don't care that an app is coded in C++, but they do care that it responds quickly. Users don't want to read a manual before using an app: they want the interface to be intuitive and self-explanatory.

To build the best possible product it is critical to understand who your users are. Unsurprisingly, that is termed being 'user-centric'.

The intersection of these two areas – building the best possible product that appeals to the biggest number of users – is the sweet spot where truly successful apps focus.

If everyone from the CEO down in your business is 100 per cent focused on delivering the best experience to users, you have the greatest chance of building traction by producing the right app. If, however, the product is not the primary focus of the CEO – perhaps the CEO is more interested in developing partnerships, or raising money, or focused on profit margins – then you will have a very hard time reaching product–market fit.

Being in constant touch with what your users want and delivering that is essential. Why? Because everything is changing so fast – people's tastes, what's popular, the competition. Unless you're focused on understanding how your users are changing, growing and evolving, you have little chance of creating a product that grows with them. And, unless this is embedded in the very DNA of your company – constantly put at the forefront by the CEO and founders – it will be hard to make that a part of what every single person does every single day at your company.

Leading the Product Vision

There are two critical roles in your app company at this stage. One is responsible for figuring out *what* product to build and the other is responsible for *how* to build it. These two people are going to be joined at the hip, so they need to work extremely well together.

The person behind the 'what' is the head of product. The person behind the 'how' is the CTO, or chief technology officer – the person in charge of building the actual software.

In many very successful tech companies – especially app-centric ones – the first head of product is usually the CEO. They have the vision about what to build, and in many cases cofound a company with someone rich in engineering experience to deliver the how. Over time, as the CEO role becomes broader and more demanding, a dedicated head-of-product role needs to be created and filled with someone entirely focused on that mission.

It's tough to lead product development. It's the role that I have been lucky to hold in a number of companies. The best product people listen to everyone's vision, assumptions and ideas, and then ensure there is a clear, data-driven process about how to build, test and roll out product improvements.

One of the key qualities of this person (besides being laser-focused on testing new product improvements and measuring their effectiveness) is the ability to say no. 'No' is a critical word to the success of any startup because it enables focus. Given limited time, money and resources, maintaining focus is the only way to get to product–market fit.

Chamath Palihapitiya is a rather outspoken product guy who was part of the team that put Facebook on the path to a billion users.[3] He explains the role of the product team very simply: testing, measuring and trying.

In order to create a killer product, all you need to do is come up with good product features, build them quickly, test them with a subset of your users, gather the data, and then, if the new feature improves one of your key metrics (how often people use the app, how much time they spend using the app or if they end up spending more money), then you roll out the feature to all your users. If the product improvement doesn't work, it must be changed or killed. It's that simple.

Since data doesn't lie, there isn't much that can go wrong with the process, unless you're not very good at (i) coming up with product improvements to test, (ii) figuring out how to measure the performance of your product improvements, or (iii) judging from the data whether a product feature is improving core metrics or making them worse.

During the one-million-dollar-app section you took the first step on this ladder of product development; you worked with paper prototypes, tested concepts with users in Starbucks, incorporated all the feedback and then even built a prototype app.

What you need to do now is adopt this process into the DNA of your company. You need to build an organisation that can follow this process of building and then constantly refining your app.

Building the A(PP) Team

To build a great app you will need a great team. What I want to outline here is the *dream team*. In an ideal world – or at least a world with enough money – you'd be able to hire all these people at an early stage. In reality, you might not be able to hire all these people now – but let's have a look at the ideal team structure so you can make sure you're on the right path.

The best way to think about which people you need in your team is by investigating which people will provide the answers to the three key questions you've been asking so far – namely:

- What is the product you're building?
- Is that product solving the right problem?
- Who are we solving the problem for (target users)?

The answer to these questions lies with the product team. Usually, one of the cofounders leads product development, as we saw earlier. If that isn't the case – because the founders are focused more on business, marketing or engineering – then it's critical to put someone in place who will own the product end to end. In these early days you need someone who can translate the founders' vision into something engineers can build and deliver as an app to your users.

When you're at the *pre*-product–market fit stage, it's unlikely that you'll have more than one person dedicated to products (ideally, you'll have a great designer working on them as well). It's only when the product takes off that you'll need to expand the product team. A good rule of thumb is to have one product manager for every six to eight engineers (so no need to worry just yet).

Depending on the type of app you're building, you may be in a different position. At Hailo we had a more complex product: a marketplace consisting of passengers and drivers. I brought in a seasoned mobile-product manager to split the workload: one of us worked on the passenger app, and one of us worked on the driver app. We needed both apps up and running to get to the stage of product–market fit. Since we raised a healthy seed round, this was a viable option for us.

If your app is less complex, stick with one product manager. It's much better to spend money on hiring engineers to focus on building a better technology for the business. The bottleneck is not usually having enough ideas about what product features to build, but rather how quickly you can deliver all the product features to your users.

As you're building your app – some other questions are going to crop up:

- What is the app going to look like?
- What is the app going to behave like?

The responsibility for answering these questions lies with the design team (which is usually part of the product team). At the one-million-dollar-app stage, you probably had the budget to engage a designer only part-time. Now, to deliver a killer product, you need to engage with a designer more closely – ideally, full-time if you can afford it.

Designers work hand in hand with the product managers to

answer the 'what' questions: What does the app look like? What does it behave like? Some people draw distinct lines between the roles of product managers, designers and even 'user experience' (or UX) experts. Personally, I find that approach amateurish and uninteresting. Saying that the product manager is responsible *only* for outlining the requirements for an app, and that a designer is responsible *only* for what an app looks like, and that a UX designer is responsible *only* for what the app behaves like is, frankly, rubbish.

In reality the product and engineering teams all work together very closely. Yes, people are definitely experts in certain areas, but at the end of the day, because we all use apps and because the domains overlap so much, people should all be encouraged to focus on the entire product experience rather than any single piece.

The designers – working with product and engineering teams – are responsible for delivering a pixel-perfect design of what the app will look like, along with wireframes to describe how a user will interact with the app. Once that happens, we arrive at a very important stage: How are we going to build the app?

Engineers or developers (I'll use the words interchangeably) are the guardians of technology. They are the ones who figure out 'how' to turn pretty designs into functioning software. They are the real makers.

Last, but certainly not least, is a critical team that ensures your app is constantly working as desired and is delivering value to your users – and answers the question: does the app work as designed?

Quality assurance (QA) – also called quality engineering (QE) – is tasked with making sure that the app your engineers are building is working as designed and specified. A large part of your QA team's responsibility is to ensure the app you're about to release to the public is bug-free – namely that there are no obvious areas where the software crashes or doesn't act as expected, or that the user interface doesn't look weird or wonky.

Let's have a look at how your app team will look in reality at this stage. One of the cofounders is going to be leading the product vision and management; the other cofounder is ideally going to be an engineer, and with your seed funding you should be shooting to get at least another app engineer (or two) on board. You probably won't be able to justify a full-time designer, but you'll need to have a great designer working with you on a weekly basis. When it comes to QA, you probably won't be able to justify someone here full-time either. In reality, making sure you have a quality app that performs well is going to be the responsibility of your one-person product department. So you'll be looking at a team of five or six people in total.

Hailo was at the other end of the spectrum: we needed to build up a bigger team (at the same time as being very tight on costs, which meant low salaries). I was brought in to head up the product development, the cofounders were taking little or no salary, we had a full-time designer (who was stretched to the limit working on the website, the passenger app and the driver app), we had a couple of iOS engineers and a couple of back-end engineers, and a couple of contractors supporting the developers, and we ended up bringing in a quality-assurance engineer because the platform was complex – and we had a lot of product development to do in order to launch! So we were more around the 10–12-person mark.

It's worth noting that in some places around the world (these are magical places) there are people who encapsulate *all* these skills – spanning product management to design to software development. I call them unicorns. Don't hold out hope that you'll ever find one – but, if you do, grab them.

Always be Shipping

Becoming a ten-million-dollar app and cementing your product–market fit is about developing product features quickly, testing them, measuring them, adopting the best ones and then repeating the process. While this sounds pretty simple to do, it does require a good understanding about software development – so let's take a moment to go through that.

The way software is developed is constantly evolving. We need to get a bit technical – but with good reason – and talk about agile software development (in the industry it's commonly referred to as just 'agile'). Agile development is methodology to design, deliver, test and deploy software in an efficient and effective way.

One way to better understand agile development is by describing what it is not. Agile is not about writing exhaustive outlines of what you would like your developers to build. It is not about planning every detail of your software development to death, trying to nail down what will be worked on weeks and months into the future. It is not about giving your development team an inflexible shopping list of features to develop and then leaving them alone to code it up.

Being agile is about delivering valuable features or improvements within a short, regular period. This period is called a 'development sprint', and it typically lasts one or two weeks. The goal of a sprint is to deliver something of value – a specific feature – from inception to delivery to your users.

Why is this pretty cool? Well, it ends up delivering usable features or improvement to your users every two weeks. That means every two weeks users are testing new features that you are measuring and evaluating and deciding whether to retain within the product. This is naturally a combination of new improvements from user feedback and new things on the product roadmap that you want to test with users.

These usable features or improvements are actually called 'user stories' – essentially a piece of functionality outlined in a very user-centric way (as in something a user can do on your app) that is designed to be developed within a 1–2-week period by a single developer. An example would be: 'As a new user of app XYZ, I would like to be able to tap a location on the map and save it to an easily accessible list of favourite locations.'

For the system to work, it is up to the product team to define the 'user stories' in detail – enough detail for the developers to be able to go ahead and implement the feature in code. That means including all the steps required, what all the individual app screens look like, the buttons, the wording, the transitions and then all the required graphics.

An agile team is defined as a group of people who can take a product feature, implement it, test it and then deploy it to users in a fully independent and autonomous fashion. This is a powerful concept, and it works really well when you have a small team. But it works equally well when you have 20, 30 or more developers, since the teams can all operate autonomously and therefore in parallel, meaning no bottlenecks in getting product improvements to users.

Working Together is Better

One thing to point out now – and I have worked on both sides of this equation and talked to numerous startups about the topic – is how critical it is that your team be located in one place. This is especially true when it comes to the product/design/development/QA family. There are many reasons for this.

Let's start with the ideal scenario. The real goal in organising your (currently small) product and development team is to give them the ability to quickly release new versions of your app – as we've seen above. It's also apparent that it's not the simplest of processes, since it's heavily dependent on effective communication.

To make sure you are constantly 'shipping', you are continually improving your code, putting it in users' hands and then getting feedback about whether it's making the app better or not. You need to seek efficiencies at every possible level.

That means having the entire product/design/development/ QA team all in one place – all within earshot of one another. Allowing one or more of your team to work remotely only makes the communication less effective. Lots of the 'soft' communication that happens when people are in the same room is lost. Lots of side conversations that happen throughout the day and night are lost. Your team is too small right now to be working in silos.

When you start approaching the hundred-million-dollar-app stage – when you start to grow at a crazy rate – communication remains absolutely critical. So the culture of communication you build now will reap even more powerful rewards in the future.

Outsource of Truth

People ask me quite often about whether they should outsource their development work. There is a relatively simple answer: no. If you want to build a leading technology company – a true billion-dollar company – then you can't outsource software development, because it's the core of your company.

That said, nothing is black or white, so let me paint a grey scenario that makes sense. No matter what happens, you need to have your key developers – and that means your head of engineering (it doesn't matter whether you call them your CTO or VP of engineering at this stage, since that comes later) – at the core of your company. Without a sufficiently competent lead engineer as a key employee you're not going to have any idea how to outsource development work anyway. How will you know what technologies they should be using? How will you know whether their architecture is sound? Blindly trusting an outside partner is not an option here.

So here's the middle ground, but it should only ever be a middle ground. Once you have a head of engineering (and ideally other engineers coming on board full time), you have someone capable of managing an external development company.

Your head of engineering or CTO should be the person making the technology and architecture decisions, and should be managing the work for this development partner. At the same time, they should be recruiting full-time engineers to join the team. If you can't do this, there is something wrong. You're either not building anything interesting or you don't have a compelling vision – or maybe you have a weak lead engineer. I have seen a lot of start-ups – with seed funding – that fall into this category. They are doomed to fail.

Other Characters in Agency Land

There are three main types of agencies associated with various stages of the product and app development process. It's worth a quick review of whether they are worth the time (and money):

FULL SERVICE AGENCIES. Startups beware: these characters were designed to bleed money out of corporates by promising to deliver not only the sun and moon but also a planet or two as well. In addition to developing brand identity, these guys will deliver websites and apps. For big-budget players like Pepsi it makes sense to outsource anything you want – but it makes no sense for a startup. If you use these guys to develop your app you will have no control over your app architecture, the quality of code and much else. Pointless if you're trying to build a technology company. And very expensive.

DEVELOPMENT AGENCIES. These are the interesting agencies. In effect they specialise in delivering software. They are usually focused on specific technology platforms, such as Ruby on Rails, PHP and Android development. They will help you on all levels: from architecture and database design, to API development, to app development and testing. They can be useful to fill in technology or skill gaps – but, as we've seen, they need strong direction. There are truly spectacular companies in the UK – such as Thoughtworks – that charge truly eye-popping day rates, but the further you voyage, the more cost-effective they can be. For example, Ukrainian agencies speak perfect English, understand all the current methodologies and are great to work with (Ciklum is one of the most popular and experienced). Polish and Romanian agencies are similarly pretty solid, and, due to their skills, language and ingenuity, have generally replaced Indian firms.

QUALITY-ASSURANCE (QA) AGENCIES. I have had mixed success with QA agencies. It's always a challenge finding talented and keen QA engineers, so you may find it attractive to work with a QA agency. You won't do it a second time. We built up a great team at Hailo. The only agencies we ended up working with were ones that were happy to provide very experienced QA engineers who were willing to work on site in our offices. They filled in gaps when we couldn't hire fast enough.

In all honesty, there are no shortcuts in building a great product-and-engineering team. You need to be able to build up a truly talented and lean team to deliver a truly loved product. If you can't recruit the right talent, then that is a reflection on the quality of your current vision, team or market opportunity. Don't ever become addicted to agencies – they are useful only as temporary stopgaps.

Chapter 15

New and Improved Version 1.0

So what does your app look like at this stage? From Step 1, 'The Million-Dollar App', you should have a prototype (version 0.1) of your app that you've used to unlock some more investment and get to this next level.

You should now be focusing on using the analytics that you're collecting to iterate and improve the existing features in your app. You should be actively talking with your users on at least a weekly – if not more frequent – basis to figure out what features you absolutely *must* add.

The features you add need to be simple, easy for users to find and unique. You need to keep developing your own unique proposition: why your app does it better than anyone else's.

To try to figure out what users really love, try killing features. Take out a feature. If users scream about it, they find it useful, so bring it back. This is the time to experiment. If your users don't scream when you remove a feature, it was probably useless. Figure out what your users love.

You need to think about the product you will launch with – your Version 1.0 (or V1.0). This is a much more solid version of your app, a refined, robust version of your prototype with a few more features. You should be confident that this version will support thousands of users happily.

This becomes an important question: you need to get your name out into the market; you need to open your app up publicly ready for anyone to try via the app stores. You can't launch with a prototype, but, at the same time, you don't want to launch with a perfect app, either. You need to get to your V1.0, which you feel is getting pretty close to product–market fit.

You need to get to this point as quickly as possible (remember, money is still very tight). You're going to need to get stuck into the following things.

DEFINE YOUR SCOPE. Get a great handle on what more you need to build. What existing features do you need to improve? What metrics need to be improved? Do you need to improve your user interface design? What about performance? Does the app need to be faster? List everything you need to deliver for a V1.0 launch. And then get to work.

TEST YOUR IDEAS. You should already have anywhere up to a thousand people with your app in their hands. You need to figure out an effective way of communicating with them frequently – at least once a week – and getting their qualitative feedback. Try also to form a group of 20 to 30 power users whom you can email or call 24/7. These people should be frank and brutal and give you detailed feedback.

Combined with your quantitative feedback – analytics – you should be in a pretty good position. You need to keep improving features, testing them and then releasing the new or improved features to all your users and measuring the results.

Usertesting.com

You're always going to be searching for more user feedback, especially the highly detailed variety. One great resource that I found very useful, especially in the early days – or whenever you want high fidelity feedback on a big new feature – is usertesting.com.

This service is wonderfully simple and cost-effective to use. It works for websites and mobile apps. You can set up user tests with any kind of demographic of people (you list your criteria and usertesting.com finds the corresponding profiles in its database of users).

You provide access to your website or app, and the users will follow your prescribed tests, answer your questions and record the entire experience on an HD webcam. The site now guarantees that results will start coming through within one hour. You have to love the Internet!

I've seen lots of startups that are afraid of negative feedback about their 'baby'. This fear is 100 per cent understandable, but it will hold you back if you give in to it. Above all, you don't want to engender a culture of defensiveness. From the very get-go, your company – and especially the product-and-development teams – should be fostering a culture of openness, inviting real people to use your app, and then inviting feedback that can be shared throughout the company to drive improvements.

One great example is Skype's mobile app. In 2013 it had more than 300 million active users across a number of mobile platforms. Jonathan Moore was in charge of mobile product there (until Hailo headhunted him to join its product team) and he introduced a great practice: get out there in front of users every single Friday and get real-world feedback. Granted, Skype has a usability lab with dedicated people working just on that (what a luxury!). As a result Skype continually tested all kinds of changes and improvements. This led to a culture of experimentation, and great improvements in not only user-satisfaction levels but also app-related revenues.

At Hailo we pursued a more cost- and time-effective approach: we dedicated a member of the product team to do UX work every other Friday on a select set of new features by using usertesting.com, and then to compile all the results, synthesise, and call a meeting with the product team to action the results.

Adopting this process will give you a great idea of whether your app will realistically be in a V1.0 launch state.

Beta Testing

Beta testing means publicly calling your app 'unfinished', or in a 'beta phase'. You're confident of its capabilities and want to get it into the hands of more people to use, but you're warning people that the service isn't perfect, and that there might be some (hopefully small) issues. It's important to be able to control how many people are using your app at this point to ensure you can deliver good performance (you won't have the server power to support massive numbers of users) and so that you can get feedback in a systematic way, so that, when you're confident your app is ready, you can get it out to even more users.

There are a couple of options available to control how many people are using your app and ensure you're running a solid beta-testing programme: Apple Enterprise Distribution and Android beta testing.

APPLE ENTERPRISE DISTRIBUTION. Apple has a special version of its distribution platform called 'Enterprise Distribution'. All you need to do to get access to it is have a registered company and upload a few details, in addition to paying a $299 fee. Once you have it enabled, you can then simply upload your app to your own secure URL, rather than to the App Store. You can then securely share that link with only the people you want to be downloading – and using – your app until you're ready to release it to everyone on the App Store. At Hailo, this allowed us to effectively test with small groups of trusted passengers and drivers, by just emailing them a link to download the beta versions of the apps.

As you might imagine, there are some third-party alternatives that help you do the same things, and these bundle in more features such as deeper analytics and messaging. HockeyApp is one great option.

ANDROID BETA TESTING. Google, on the other hand, launched a 'beta testing and staged release' feature directly into its operating system at its I/O (the I/O here is a bit of a geeky computer term and stands for input/output) conference in 2013. In fact it's a pretty ingenious feature: it allows you to upload a test version of your Android app to Google Play and automatically prompt special users – who have beta-testing status – to download and use the test app. You can hand out this privilege on a user-by-user basis – or even select a fixed percentage of your Android user base. Then you can monitor the results – for, say, a 5 per cent sample – before releasing it more broadly. Genius!

You also have another option. By now, if you've done your job right, you have a decent list of people who have shown interest in your app by submitting their emails on your website. This is a great list of people you can begin testing with: they are keen, want your app and have a likelihood of reporting issues and suggesting feedback. Launchrock.com, which we talked about earlier, even has an inbuilt feature to simply email all these signed-up people with a single click. Hey presto! Your beta-testing programme is launched.

A Chelsea Housewife in Two Taps

One of the clever things that we did at Hailo was to very carefully select and communicate with our beta-testing community in London. Picking your target users is critical: you wouldn't test a Square register solution with a gamer, and you wouldn't test the Uber private-limousine app with someone on a low salary who doesn't frequent the city much.

Given that Hailo needed to be designed for your average smartphone user (in order to maximise adoption), we couldn't afford to have a product that had high barriers to adoption or first use. In fact, we designed the app with precisely the opposite intention in mind: it had to be intuitive for a first-time smartphone user. Hailing taxis is, after all, a mass-market proposition. Consequently, with our beta testing we wanted to target avid taxi users, who would use taxis several times a day, not be particularly price-sensitive, have the time and interest to actually send us feedback and be not necessarily models of early technology adopters. Thanks to some friends who lived in the rather affluent part of west London, Chelsea, we were able to email out a link to download the app, combined with a lot of free credits, and we were able to quickly have a number of real-world people use the app (with their real credit-card details) to hail our few hundred beta drivers who promised to accept every single hail and give us feedback about the experience as well.

As it turned out, this was the best thing we could have done. We were increasingly swamped with requests for their friends to get the app (remember, it wasn't available on the App Store just yet). And we obliged. Given how much the community talked, we were able to get hundreds of people – including friends, family and banker husbands – who habitually took taxis to and from work every day. The clever beta drivers cottoned on to what was happening and started hanging out in the financial district every afternoon as well.

Feedback Loops

This beta-testing period was invaluable. As a result of the feedback we received, we made a variety of changes to the app: how the taxi driver's confirmation message and user interface appeared to users; how we showed the user location on the map to drivers; how easily the driver could call the customer to confirm/clarify their pickup location (it turned out to be very important to enable this communications loop to circumvent early location-accuracy issues) and a number of smaller usability factors.

The testing also gave us the objective confidence, from hundreds of real people, that our service actually worked, was fun to use and was as ready as it ever would be for prime time.

App Store Ratings plus Comments

For most apps at this stage it probably makes sense to release to the Apple App Store or Google Play (depending on which OS you have chosen), unless you really need to be careful or secretive. In this case you should be vigilant about ratings and reviews – they can be a great resource of feedback while you're shooting for product–market fit and the better overall rating your app has, the better it will do in search results. Make sure that you have everything in place in terms of app-store optimisation outlined earlier when we spoke of the million-dollar app.

Make sure you monitor your ratings on a daily basis. On Google Play you can respond to users with issues (it's critical to win people back in the early days); Apple keeps promising to add this feature to its App Store but, at the time of writing, there is no clear timeline.

Your app-store ratings definitely help persuade people to

download your app because the impact of an app with hundreds, or even thousands, of four- and five-star reviews makes the download decision easier. A larger number of better reviews improves your ranking in the app-store search results. All of these factors bolster your brand – and the level of trust users have in your app. It's important down the line when you ask users to pay for services: they will enter their credit-card number or make an in-app purchase only if they trust you.

Make sure that you create a DNA of customer support from the very first day in your company. The cofounders should be checking the reviews daily and personally responding to customers. You should never lose touch with your users, since they control the ultimate success of your company.

And one of the tricks of the trade: ensure that you get the most, and highest-quality, reviews from your users. First, prompt users to rate your app after they have had a great experience (e.g. when they finish at a high level in your game, receive a delivery from your e-commerce app or complete a successful taxi ride). You can track how many times you prompt people via your analytics solution – and you can also track how many ratings you're getting on a daily basis with app-store analytics tools such as AppAnnie or Distimo.

And, second, keep your ratings fresh. While you shouldn't annoy your users, do make sure that you prompt them to update their rating of your app every six months or so. Again, this should be an automatic prompt that is built into your app. Make sure that you make it as simple and pleasurable for your users to tell everyone how great an experience they are having with your app.

Delivering Delight

A great company loves to delight its customers. It's one of those things that should be built into the DNA of the company – and, naturally, into the product-development process. Not enough apps focus on delight; they don't realise the amazing benefits it can drive, such as repeat usage, word of mouth and ratings on app stores.

It can be tricky to measure it. At Hailo we found great success in tracking sentiment via social media: we received a lot of very happy user comments via Twitter and Facebook, and many bloggers would take screenshots of the app and post them. We made sure we kept track of this and measured all the mentions we were getting – and what events were triggering the mentions.

Further down the line, we started using the Net Promoter Score (NPS). This is a very simple survey that you send out to a representative sample of your users to measure how likely they are to recommend your app to a friend. On a 1–10 scale (with 10 meaning highly likely to recommend), people who give you a 9 or 10 are called promoters; those who score 7 or 8 are neutral; and everyone else is a detractor. Your goal is to make your average NPS across all users as high as possible; it's an ongoing measure of whether you have product–market fit.

Delight directly affects *retention* of your users. The best apps (and services, companies, etc. in general) can grow only if they are 'net-adding' users, i.e. they are adding more users than they are losing. That's definitely not the easiest thing to do – and you'll find that you'll need to employ numerous simultaneous strategies to make it work.

So what are the important things to get right in order to delight people?

Design is one. If you've used the Hailo app you'll have noticed there is a little blue man in the middle of the app who denotes your current location. In the very first version of Hailo, this little blue guy (whose nickname is Barty – named after a summer intern) was just a blue pin. The pin was clear, simple and well recognised. One of our designers – a rather emo-looking fellow, who plays in a band, loves tattoos and used to make video games at Electronic – didn't think it was good enough.

One day he told me he wasn't happy with this impersonal feel of our passenger app; he said it was lacking something. I agreed. So I threw the challenge right back at him. 'You're the designer, Dave. You have the rest of the week to come up with something brilliant. After that we need to ship it.' After hacking for a few nights, he waltzed onto the boat on Friday morning and thrust an illustration before me of a little blue man, sitting on top of a location pin, arm raised as if he was hailing a taxi. The head of design was equally impressed. We all knew he needed to be in the app. Today, Barty is continually commented on by users and by the media, and has even become a central character in our marketing campaigns.

FLIPPING DELIGHT. Flipboard is amazing. This app allows you to automatically generate social and personal magazines. Marcos Weskamp, the head of design at Flipboard, created an original, simple and fun-to-use app. His design team really understands the use case – of using an iPad like a magazine, quickly, simply and visually flipping through lots of content and making it easy to consume and share. Even a few years after its launch, very few apps had focused on delivering such a great user experience, thus allowing it to remain the undisputed leader in its space.

ANGRY BIRDS. Gaming is cheating a bit, since these guys are in the business of delivering delight day in and day out. That said, Angry Birds took that to a completely new level. From the very first interaction with the game, you can see it was optimised for a touchscreen device. The bouncing of the birds into your slingshot, the way the trajectory of your next 'shot' is overlaid, even the character noises and soundtrack – every element of the game is designed to educate, entice, excite and exhilarate you.

Communicating with users

One of the unspoken secrets of great apps is their customer support. If you engage with users then you are going to help drive all kinds of other metrics – especially all the social ones. So it's probably a good idea to set up a decent support solution at this point.

We knew during the early days of Hailo – operating a marketplace model with passengers and drivers and dealing with real payments from Day One – that we'd have to keep everyone happy. That meant dedicated customer-support people from the outset. We invested in a SaaS solution from desk.com early on, which managed all the support emails and enquiries, allowed us to centrally manage and reply to them and allowed us to measure how quickly and effectively we could reply to our customers' issues. It also allowed us to easily categorise all the support issues, so we could then see what parts of the app or service needed to be improved. It was powerful and easy to use. It also grew with us, supporting a team of customer-support staff without an issue.

The customer is always right

Lots of our users would go to social media to complain about all kinds of things that went awry during the early days (as you do). Luckily, our social-media guru Chris was on board during the very early days of Hailo as well (if you aren't lucky enough to have a dedicated person dealing with this, then it's going to be yet another thing you as a founder need to be on top of). By adopting a philosophy of being very open and communicative about issues and attempting to solve them all publicly and expediently, we built up even more fans.

We built a nice system of Hailo credits to reimburse users for any issues they had, and also as a thank-you to our best users. Don't forget about your best users – no one likes to be taken for granted!

One brilliant example of why it's a good idea to take customer service seriously was an incident we had one Friday evening when we were working late on the boat. A tweet was tweeted – along with an email sent to support@hailocab.com – from a woman who had had her credit card pre-authorised several times for £100 instead of the usual £20 (due to a software bug), which prevented her from withdrawing any money that evening. She was naturally pretty upset. Given that we couldn't do anything from a system point of view on a Friday night (it would take a few days to sort it out with the bank), Jay, our CEO, jumped in a cab to go directly to the customer's house with £100 in cash, and a bottle of wine to personally apologise for the inconvenience. What started as a disaster ended with an emotionally thankful tweet from the affected customer. A great turnaround.

Measuring Product–Market Fit

Let's briefly go back to the metric mentioned at the beginning of this section: '... achieving product–market fit requires at least 40 per cent of users saying they would be "very disappointed" without your product.'

How do you practically go about measuring this? One strategy mentioned earlier is to use a survey, one like the NPS that you send out to your user base. This should give you a pretty good indication, though you will tend to capture either the happiest users (who are happy to reply to your email) or the least happy (who are happy to complain). You typically won't get a high percentage who reply.

The other, more powerful and objective way is going to be

through your analytics. Are more than 40 per cent of your users coming back often? Are users spending more time on your app? Are users actually referring other users to download and use your app? All these things can be measured.

In the next chapter we take metrics and analytics to the next level. Not only do we now need to focus on ensuring product–market fit, but we need to start measuring the effectiveness of the growth and revenue engines.

Chapter 16

The Metrics of Success

'Build something users love, and spend less than you make'
#BILLIONDOLLARAPP

You need to measure things to improve them. I get very excited about being able to objectively measure how much users love an app that I have helped build. There is something great about creating a positive feedback loop that constantly drives improvements – improvements that first and foremost enhance the experience for users, and, as a by-product, improve the performance of your business.

Another simple piece of advice from Y Combinator's Paul Graham takes us to the next level of measurement: 'Build something users love, and spend less than you make.'[1]

To achieve this you first need not only to be able to continually measure how much users love your app but also to start building a picture around the costs – and revenue – driven by your users. This is the core of building your growth and revenue engines.

Earlier, we used Dave McClure's pirate-sounding metrics, AARRR, as a starting point to measure the performance of your prototype app. Now, it's time to take that one level deeper: to flush out the critical metrics that will underpin your app as it begins to grow. What is also important is to clearly communicate what success means to everyone in your team. And, as the team grows, you need to be able to communicate who is responsible for delivering each one of those metrics.

Riding the Customer Lifecycle

The app world is a sea of metrics. Focus is always going to be key to achieving success – especially when it comes to metrics.

Let's do a quick recap of the metrics that we looked at earlier – AARRR:

- **A**cquisition
- **A**ctivation
- **R**etention
- **R**eferral
- **R**evenue

Throughout this stage – as you're approaching product–market fit – you need to think about each step of the customer lifecycle as a funnel. You need to think about moving your users from the first stages of the funnel (acquisition, activation) through to the later stages (retention, referral and, ultimately, revenue). As you move users through the funnel, you will be moving them from being of lower value to being of much higher value.

Start by defining your own customer-lifecycle funnel, complete with stages and conversion goals, in a table like the one opposite. Once you have completed that, ensure that you can measure the conversion rate from one stage to the next (via your analytics solution, be that Google Analytics or Mixpanel). With a baseline in place, you can now focus on making product improvements to enhance each one of those conversion rates.

We are building on the table introduced earlier in Step 1, so this should look familiar. But now we're going to add in the conversion rates. I've included some very general ones in the table. They will vary a lot based on your app, market and business model, but they generally make sense.

Let's use the example of an e-commerce app.

Stage	Conversion goal	Conversion rate*
Acquisition	User downloads app	100%
Acquisition	User launches app – either taps any button – or session time > 10s	90%
Activation	User creates an account	70%
Activation	User adds credit-card details (required for fare payment)	50%
Activation	User adds item to shopping basket	40%
Retention	Opens app from email/push notification alert	50%
Retention	Opens app 2+ times in the first week	10%
Retention	Opens app 5+ times in the first month	10%
Referral	Refer 1+ users who download app	3%
Referral	Refer 1+ users who activate	2%
Referral	Refer 1+ users who generate revenue	1%
Revenue	User makes purchase < $5	3%
Revenue	User makes purchase >$5 and <$25	2%
Revenue	User makes purchase > $25	1%

* The numbers are purely illustrative and will vary depending on your app design, sector, user base and numerous other factors.

So, once you have mapped out these user-conversion funnels, you're in a good position to understand how your app works from a user-value point of view. You now have a clear set of user journeys that you need to move your users through to make your business work; you have clear visibility of the metrics; and you know how to measure conversion rates from one level to another.

The hard part is figuring out how you need to change your app to drive the numbers in the right direction. Your mission is to optimise this entire customer-lifecycle funnel, taking into account user happiness – you want your users to enjoy going through every step of this process. You need to make it as simple and enjoyable as possible to part with their money. Second, you will also need to optimise the funnel for maximum revenue and lowest possible cost (thus driving your profitability).

While all apps are created different – and your conversions are going to vary quite a bit – you should be shooting to activate around 80 or 90 per cent of your users. The typical conversion for users who make a purchase on any given day on an e-commerce app is around 4 or 5 per cent – but many apps are able to do a lot better. In a later chapter we dive deeper into conversion rates you can shoot for around referral.

A Few More Killer Metrics

Before we get to the mechanics of it, we need to add a handful of key metrics to the mix.

AVERAGE TRANSACTION VALUE (ATV). This is a very important metric to keep your eye on. This is the average amount that a user spends via your app. In a gaming app it will most likely be via an in-app purchase and hover around the $0.99 mark. In an e-commerce or marketplace app it will be the value of the transaction (such as a taxi fare) or the basket size. For a SaaS or enterprise company this is going to be the monthly recurring fee that you can charge for users of your service. This is something you will want to validate very quickly because you need to be able to estimate how much your users are happy to spend with you during a given interaction.

ANNUAL REVENUE PER USER (ARPU). Although this is sometimes defined as the average revenue per user, I don't find that precise enough. It's helpful to associate the revenue with a time period. You will need to adapt the time period that will make sense for your business, since many apps will end up using monthly revenue per user. Right now you're probably going to have to estimate this – but it will prove to be very useful to model your future revenues.

LIFETIME VALUE (LTV). The final killer metric here is lifetime value. This is basically your annual revenue per user multiplied by how long a user stays with you. It's evident from all these metrics that you want to ensure that a user stays with you for the longest time possible, so that you can maximise their revenue potential.

For the moment these are the top metrics you need to keep in mind. Through the process of building an app people love, you're going to be hunting for new ways to attract users who deliver the highest lifetime value and the lowest customer acquisition cost. The philosophy is simple and logical, but the practice is a lot harder.

Consensual Metrics

Just having your own metrics in place is not enough. Metrics should be clear enough so that they become a rallying point for the entire company. Everyone in the team should know which metrics their work can affect – and should feel empowered to improve those metrics. To that end, Jack Dorsey, CEO of Square, actively shares all the core metrics of his app with everyone in the company on a weekly basis. Naturally, this can be pretty sensitive, but Square considers employee trust as massively important to its corporate culture.

At Hailo, we similarly shared critical metrics with the entire development and product teams. This ensured that we were all critically aware of what was going well and the areas that we needed to focus on. Sharing metrics is critical to building visibility – and trust – within an organisation.

AGREEING ON METRICS. Companies often have arguments about metrics because they haven't spent enough time discussing and agreeing on definitions. Metrics are not going to be helpful unless there is buy-in from all the people who are affected by that metric. So, when you're going through the process of deciding your customer lifecycle stages and the conversion metrics, make sure you bring everyone along for the ride.

TRUSTING YOUR METRICS. It is important that all teams – whether development, product, marketing or operations – understand the data that goes into defining a metric. I have found that, if people don't fully understand the source of their metrics, they can lose faith in the accuracy of the metric itself. This is a bad situation. Once any confusion is removed at this point, there is much more belief in the power of the metrics themselves. It's a good idea to periodically compare what your analytics platform is saying to the source data pulled from your own databases to validate that what you *think* you are measuring is actually *what* you are measuring.

You Can Measure Sh*t Better

So it's clear that measuring stuff is important. It's time to learn to love numbers if you don't already. For those of you with a bit of a phobia of data, let me start off with some good news: analytics software has never been more friendly to use, simple to install and cost-effective to run.

We've discussed settling on the key metrics by which you're going to measure your app's success. Great. Now you need to access those numbers in a way that's easy to view and consume – and also in a way that ensures you trust them, since you're going to be making life-changing decisions based on them.

In a nutshell, analytics software gives you insight into who is using your website or app, and what those users are doing. Are users landing on your homepage and leaving in three seconds? Or are they landing on your homepage, clicking through seven pages and then making a purchase? Analytics software tracks every 'event' that happens on your website or app, then collects that information and presents it in a nice visual interface.

This interface then allows you to read through a variety of metrics – such as how many users are using your app – and then view them on an hourly, daily, monthly basis. The interface also allows you to see what users from the United States are doing on your app, and see if that behaviour differs from users in the UK. You can also see trends in behaviour. Are people using my app for longer this month, compared with last month? Or are they using it less? The wealth of information available to you via analytics illustrates why having a clear outline of what success means to you – and your app – is critical. Analytics by itself is powerful, but it is also very broad, and you need a clear set of questions you'd like it to answer to avoid getting overwhelmed by analytics software.

Previously, analytics solutions shied away from providing information that drilled down to the individual user level (mainly for privacy reasons). But new startups such as Mixpanel are pushing the boundaries, and can be massively more useful, because not only do they deliver all the analytics you'd expect from a best-of-breed solution but they also allow you to see how each individual user behaves on your website or app. This is a fantastic development, because no longer are you forced to treat users as groups, or segments, and now you can reach out to them as individuals. You can see precisely what *each individual user* is doing on your app, and you can tailor the experience directly to him or her. For example, you can tailor alerts, messages, even promotions based on their individual behaviour, and then monitor their individual responses. This wasn't possible before in an off-the-shelf analytics solution.

Analytics are going to be your best friend – a friend who doesn't give their opinions but rather tells you the way it is.

Rookie Mistakes

PUTTING IN ANALYTICS TOO LATE. Installing an analytics solution before you launch your V1.0 is the golden rule – hopefully you've followed my advice back in the section on the million-dollar app. Your website or app is nothing without analytics – you're going to be flying blind with no idea about how many people are even using your app – so make sure you get the code in there immediately. There's no debate on this one.

RELYING ON A SINGLE ANALYTICS SOLUTION. As we saw earlier, it's very helpful, especially in the early days, to put in two analytics solutions rather than one. With recent performance increases, having two (or three) solutions in either your app or website is not going to either slow down or overcomplicate things. And, given how much of a pain it is to transfer data between systems, there is very little cost.

▶

NOT ATTRIBUTING MARKETING OR REFERRAL SOURCE. This is perhaps the biggest mistake you can make. From the start, you really want to know the source of a user. Did a new user come to your app organically? Did they come from your website? Another app? Or from a PR or partner campaign? Facebook or Twitter, perhaps? Or a specific, paid-for marketing or affiliate campaign? You can get pretty much 99 per cent accurate attribution these days for apps. You can do it by working with a partner such as mobileapptracking.com. So do it from Day One. From the get-go you will have visibility and the ability to invest super-efficiently.

NOT PLUGGING IN REVENUE METRICS. Useful metrics such as average transaction value (ATV), annual revenue per user (ARPU) and lifetime value (LTV) can be easily calculated by a number of leading analytics solutions. It's definitely worth investing the time early on to track this. Otherwise you tend to make excuses, and it can become hard to calculate (and often less accurate) via spreadsheets.

Check Out These Four Analytics Tools

There are a variety of solutions on the market, and they all have their sweet spots, so it's worth knowing what's out there. I've listed my personal favourites below.

GOOGLE ANALYTICS. This is a solid solution. Get it set up and working with your app (a relatively simple process), and also make sure that it's installed on your website. It will enable you to track pretty much everything that you need to. Some parts of the interface are a bit complex and confusing (and take a bit of learning), but there is no downside to putting this solution in place. Specific features that I like are: real-time analytics (so you can see what is happening on site during busy periods), great annotations (so that you can drop in comments about when you released new features and track the changes in your analytics), superb performance and pretty solid funnel analysis (so you can robustly track users throughout the various different lifecycle stages in your app).

FLURRY. This is a free, solid solution and definitely an OK place to start, but you will quickly find yourself demanding a lot more than it can provide. There is little if any segmentation (very weak compared with Mixpanel – see below) and it doesn't really process information that fast (it does support hundreds of millions of users, though), but still it's worth putting in place as a basic solution. It's not a bad idea to start simple, master your current analytics solution and then upgrade when you outgrow it.

MIXPANEL. Boom! I have to say that when I first started playing with this solution I was filled with shock, awe and excitement. Despite being a fairly new company (founded in 2010 and from Y Combinator), these guys are killing it. Not only are they a standard in terms of mobile and Web analytics, they do very advanced calculations around cohort analysis, LTV and a series of other advanced – and very helpful – calculations. Segmentation is another super-strong area for them: you can visualise and filter all your data by all kinds of segments, from profile information that you get from your users, to all kinds of behavioural data (how many times someone has done or failed to do something on your app), to any other piece of data you have about a user. As a bonus point, the particularly interesting (and unique) feature that Mixpanel has is its ability to show you all behaviours/actions on a per-user basis. Why's that interesting? It already has the ability to send emails, SMS messages and mobile push notifications via its dashboard (this is very much invading the playground of CRM, or customer-relationship management, solutions, but more on this later). The solution is crisp, powerful and rather ingenious. At scale it can get pricey, but worth checking out to see if it's right for you.

LOCALYTICS. These guys are also pretty great and probably most similar to Mixpanel, except with a smaller feature set. They lack the power of doing LTV (or complex calculations) on historic data, and the platform is definitely slower. They also lack all the cool CRM features.

This is not meant to be an exhaustive list, and, frankly, you could spend a lifetime reviewing them all. But without a doubt you can do everything you need at this stage with these analytics solutions.

I'm going to dive further into data-analytics tools in the hundred-million-dollar app section, but this time with a focus on what is now commonly termed 'big data'. I'll go into tools that will help you visualise and get insight out of all the information you have, so that you can make better business decisions. We'll check out how solutions such as QlikView help you get the data in a form you need to help the company to make faster, better decisions, and we'll look at how to get buy-in from the bigger team.

Chapter 17

Getting Your Growth On

Marketing is about two things: acquiring users and retaining users. At this stage we're going to focus mainly on building up a solid user-acquisition strategy – your growth engine – so that you can reliably acquire valuable users (ones who come back, refer other users and generate revenue) for a known (and efficient) cost.

Your user-acquisition strategy is going to be focused on experimentation and validation. You need to go out there and find the most efficient channels to get users downloading your app, and at the same time you need to test what campaigns – messaging, wording, imagery, propositions – are going to get those channels to perform the actions you want.

Since you're going to be constrained by time, money and resources, I'll start with some pretty logical channels, the ones that are high-volume, high-conversion and low-cost. For each channel you should test a variety of campaigns (which I will go through below) and then start putting together a picture of which channel–campaign combination yields the most users who deliver on your desired conversion goals for the lowest price.

At that point you keep iterating – fast – focusing about 80 per cent of your time testing product improvements and features to improve the conversion rates on your main metrics. The remaining 20 per cent of your time should be spent thinking about brand-new, unique features that make users love your app even more.

This process is one that you'll be repeating throughout the life of your app. Your goal is to get to a point where you have a great idea about the channel–campaign mix that delivers users at a cost well under the lifetime value they generate on your app. When you get to this position, you'll have a fantastically profitable business.

Getting Emotional with Campaigns

For users to love your app, you need to be able to actively make them feel something. Your goal is to get users to love – or even hate – your app. Death is when they don't feel anything strong about you and you become someone stuck in the zone of indifference. Red Bull has achieved that well: it formulated its taste so that 50 per cent of people would more than like it: they would *love* it. And it actively didn't care about the rest (the company is worth more than $7 billion).

Snapchat has been able to do something similar. Snapchat is a simple concept – messages that disappear after ten seconds. You either get it or you don't. People who get it use it feverishly. People who don't talk about it, complain about it, and rant about teens using it for sexting. Both groups of people generate massive interest and PR, and that has led to huge adoption. If Snapchat were to placate the naysayers, and change the app to make it more attractive to everyone, it would lose a huge chunk of its organic PR and probably alienate its core user base who get it anyway.

You need to discover the triggers that drive people to want to download, use and talk about your app. Start by finding the key-words, themes, phrases and images that excite people about your app. This will become your lexicon – a word universe – that you can use throughout your marketing campaigns.

Start by making a few lists of the following things:

- The top five emotions you want your app to elicit
- The top fifty words that describe your app
- The top fifty words that describe your brand
- The top customer needs your app satisfies and benefits it delivers
- The top problems your app solves
- The top fifty words that describe your competitors' apps
- The top fifty words that describe your competitors' brands.

At Hailo, we found that words like 'hail', 'instant', 'whenever', 'wherever', 'automatic', 'two taps on your smartphone' and 'credit-card payment' were all terms that people responded to incredibly well. We distilled a lexicon, and then used it across our website, app, PR and marketing material.

It may also help to do a similar exercise and list the top people/demographics associated with your app, images of other products associated with it, or images of associated problems or solutions you provide.

The end result is to generate clear themes, messages and calls to action that you can test in campaigns across all of your customer-acquisition channels, to see which ones attract the most valuable users.

Testing Channels

Attracting large numbers of users to download your app and then keep using it is hard work.

Yes, there are a handful of apps that go 'viral'. Suddenly, everyone seems to be downloading them despite their having no discernible marketing campaign (we'll look later at what 'viral' really means).

Those outliers aside, there are a number of tactics that leading app companies put in place from the start to ensure they are attracting users in a systematic and cost-effective way. These tactics invariably come from experience, meaning startups typically miss a few of these entirely, adopt others too late, or completely fail to get the groundwork in place before investing in user acquisition.

We put a number of these basic tactics in place in 'Step 1: The Million-Dollar App': ensuring you have a responsive website with a clear call to download your app; core SEO (search-engine optimisation) elements in place on your website; ASO (app-store optimisation) in place; and then a basic understanding of PPD (pay-per-download) advertising. We also talked about leveraging social features in your app, empowering people to more easily share your app and broadcast it to social-media channels.

We also talked about publicity. That's going to be one of the biggest low-cost channels you can pursue at this point. Reaching out to bloggers, app review sites and similar online channels is going to give you some good exposure, especially if your app is getting some solid product–market fit. It's going to be a pretty tedious and time-consuming exercise, but it will lay the groundwork for a growing network of people keeping an eye on the growth of your app.

Now it's time to talk about experimenting with some bigger channels.

A New World of Big Channels

We live in a world of new, massive platforms. There already are a large number of players in the market who reach anywhere between 100 million and 1 billion people. That is a distribution mechanism like none other in history – and that presents a huge opportunity for your app.

Let's have a look at some of them.

In the search market there are well-known players such as Google, with massive worldwide reach; Baidu, which is the leading search engine in China; Yandex, which is the king of Russian search; and such engines as Yahoo! and Bing. All are optimised for desktop and mobile marketing.

On the social level, Facebook has more than a billion active users and a very powerful mobile-advertising platform; Twitter has hundreds of millions of users and a growing mobile-advertising platform; Tencent QQ is the undisputed social leader in China with around a billion active users across its services as well; LinkedIn also provides a great platform to reach a professional audience with more than 200 million users.

We already know about Apple's App Store and Google Play, which provide massive reach as well. Getting featured on these sites is a massive boost to downloads, so using any means necessary to get featured will translate into downloads.

Media channels such as video (YouTube), photos (Pinterest, Instagram) and blogs (Tumblr) are all proving to be reliable, high-volume user-acquisition channels. Your challenge will be to see how to make those channels convert to valuable users.

Messaging platforms are not yet being widely used as user-acquisition channels for other apps, but, as they keep increasing their reach, this will be a monetisation for the likes of WhatsApp, Snapchat, WeChat (China), Line and even Skype.

In terms of local advertising, it's also worth exploring what you can do with players such as Groupon and Yelp, each of whom have more than 100 million users.

Other channels to think about include Amazon (and its own app platform) and PayPal, which often partners with companies to help promote various apps and services. Players such as Alibaba, Rakuten, Sina, NHN, Yahoo! Japan and SoftBank are all potential distribution partners. SoftBank was integral to the adoption of the Clash of Clans app in Japan, and eventually bought the company behind it for over $1.5 billion.

It's All About Attribution

The first step to figuring out which channels are working for you is making sure that you know which channel is responsible for driving a user to download your app. You basically want to identify the source – the website, the app store, the search term, the referral – that led every single one of your users to download your app, complete an action or create an account. This is called source attribution.

This is something you *need* to be able to track from the very beginning. Why? If you know what channels drive downloads – and, more importantly, what channels drive users who use your app frequently, or spend a lot of money on your app – then you can focus on those channels. Since you have limited time, money and resources, focus is critical. And focus will allow you to optimise your spend on the users who are driving value in your app.

Just imagine if you're *not* doing this. You've experimented with a variety of channels; you're seeing a wide variety of results – fantastically engaged users, others who open the app only once. How do you know which channel is bringing you which user? Remember that it often takes a bit of time to compel users to spend money with you. If that process takes a few weeks, how are you ever going to know which channels or sources worked for you if you didn't record them?

So how does the attribution work? Let's start with a simple example: how ad tracking and attribution works on the Internet. Then we can extend those concepts to the slightly more complicated world of mobile apps.

Imagine you're looking for a new pair of sunglasses. You type your search into Google, get a page of search results and see an ad promising 'All sunglasses 50 per cent off for the next 24 hrs'. You click on the ad and are redirected to a website selling sunglasses. After browsing around the site for a while, you find a pair you like, and buy them. And, sure enough you saved 50 per cent – pretty neat.

So, throughout this experience, a number of things have happened.

- When you visited Google and searched for sunglasses, Google gathered a few pieces of information to uniquely identify you: the type and model of your computer, the type and version of your Web browser, the internet service provider you used to connect to the Internet with and a large number of other parameters.
- Google also dropped a small file – called a cookie – into your Web browser, which uniquely identifies you to Google. If you already have a cookie, and Google recognises it, then it knows you've visited Google before and has information about what you searched for.
- When you clicked on the sunglasses ad, Google recorded that too, and knows that you went to a specific website from an ad in its search results. It then shares this information with the merchant operating the website.
- Since you purchased the sunglasses in the end – a win for the advertising website – it knows how you arrived at its website (via the

Google ad you clicked on). The merchant also knows specifically which ad you clicked on (i.e. which offer you responded to) and it also knows that you searched via, say, a laptop computer while using the Safari Web browser.

As a user, I am completely unaware that all this information is being collected about me (it's not necessarily a bad thing). But it is great news for the person running the sunglasses website, because they were able to communicate an offer that appealed to me and I made a purchase.

How about mobile attribution?

The app world is slightly different. Smartphone users are not the same as Internet users accessing the Web via a browser. On a smartphone, users are constantly hopping between apps. Since these apps are separate and distinct (they are all created by different developers) you can't just drop a cookie on the user's smartphone to track their app usage.

So what can you do?

Naturally, given how valuable mobile advertising is, brilliant businesses have come up with clever solutions. Here's a top-level summary of what happens.

- You create a tracking link for your mobile advertisement (this is unique to the advertising partner and program).
- When a user interested in your app clicks that ad, a unique fingerprint is generated that is tied to identifying information about that user's smartphone.
- The user then goes to an app store and installs your app.
- Then a piece of code you installed in your app regenerates this fingerprint (based on the identifying information about that user's smartphone) and sends it to a tracking server.

- The tracking server then matches the two fingerprints, and then matches the one sent from the smartphone to the one generated when a user clicked on the ad.

One of the top solutions on the market that adopts this rather cunning approach is called Mobile App Tracking[1] (zippy name!). I've included more information on mybilliondollarapp.com for those interested in more of the fun details.[2]

Attribution for referrals

One powerful use of a mobile attribution platform is for referrals. A lot of startups underestimate the power of referrals early on, and exacerbate the problem by not measuring it properly. If you really want to drive referrals, make sure you can measure them from the start.

In additional to using systems such as Mobile App Tracking, there are a number of other simple ways to track – and drive – referrals.

Let's start with the simplest one: redemption or referral codes. Generate simple, short and unique codes that users can send to other users. If you design your campaign carefully, and ensure there is sufficient motivation for both the 'inviter' to send the codes (e.g. they earn credits or money for each one redeemed and not just shared – remember that!) and sufficient incentive for users to input the codes (get £5 off your first taxi ride with Hailo), then you will encourage existing users to refer your app and new users to redeem codes to use it.

The codes make it very easy to then check on the back-end (in your user-management system) which users joined because someone else invited them. You can then figure out who that user is (all your codes should be unique for this purpose). You can also see which existing users are responsible for inviting the most new users, and focus your efforts on seducing them even more, and rewarding them for their effort (or figuring out how to make them even bigger advocates of your app). If you don't have a way to measure this, you won't know what your success rate is and you won't be able to improve it.

So the above referral code system works well if the referred user actually puts in the code. If they don't, you can't track the referral. That is unless you use a system like Mobile App Tracking in conjunction with the referral code. If you embed in the email or SMS a tracking link that is used to share the referral code, then you can still track the referred user.

It's a pretty powerful system and it's worth getting your head around.

So why is this clever?

It's worth investing the time in tracking all your advertising and referral in this way because few startups do it well in the early stages. That means that you can take advantage of the situation and grow faster, with a more efficient marketing spend early on.

Let's use a real-world example from Uber – the chauffeur app. It used very detailed referral tracking from the onset. What did this do for Uber? Detailed tracking allowed Uber to see that some users were referring a higher calibre of user; there was a clear segment of high spenders referring other high spenders. So what did Uber do in response? It increased the value of the referral codes for these users. This increased conversion substantially. Uber consistently offers $20 referrals for some users – and $50 for others. This drove better conversions and faster adoption early on – and was replicated across all its cities.

It doesn't cost you anything to get a powerful tracking solution like this in place – so why not do it? It can only give you better visibility.

Chapter 18

Dollars in the Door

Now that you've been testing your growth engine, you should also have been testing your ability to generate revenue. Let's have a look – based on your business model – where you should be in terms of generating revenue.

GAMING. You have no excuses – you should be making solid revenues, have a solid idea of ATV and ARPU and be able to model your customer LTV.

E-COMMERCE. You also have no excuses. You should be pushing transactions through and selling your products or services and have a similarly strong grasp of your revenue metrics.

MARKETPLACE. You also should now have enough traffic through your app to show that you have priced your service correctly. You're probably more focused on the ATV than on the ARPU or LTV, which will become clearer as the marketplace develops.

SOFTWARE AS A SERVICE (SAAS). You also should have a good idea of the amount users are willing to pay for your service and you should be generating revenue.

ENTERPRISE. You should have a handful of paying clients and you should have been able to validate price points and average transaction value, and, hopefully, a decent estimation of your annual revenue per user.

Let's have a look at what the best startups are achieving at this point. According to Ash Fontana, a hacker from AngelList (and who, coincidentally, went to my high school back in Australia) has compiled this list of benchmarks[1] – based on data from thousands of startups on the AngelList fundraising site.

To raise $1 million at a decent valuation of around $4–5 million you need:

- Great product
- Great team
- Great traction

At this point we've put together a great founding team, and we've been iterating towards a great product that fits the market. The last element is traction – and that is evidenced by your metrics. According to Fontana, these are the numbers you will need to show to secure further investment:

- If you're an **e-commerce** (or **gaming**) site or app you should be generating $50,000 per month
- If you're a **marketplace** you should also be generating $50,000 per month
- If you're building a **consumer audience** app you should have at least 100,000 downloads

- If you're an **enterprise** company you want 1,000 paying users at $10 per user per month.

These are pretty tough goals to hit, but that is the bar for professional investors today. If we look at another source, Dave McClure (the man behind the 500 Startups), we see that the numbers are even tougher. According to McClure, you should be hitting the following numbers:[2]

- If you're an **e-commerce** or **marketplace** you should be generating around $100,000 per month (or north of $1 million in annual revenue)
- If you're shooting to build a **consumer audience** you should have 10 million downloads, and at least 1 million active users
- Ideally, you should demonstrating 100 per cent year-on-year **growth**.

It is true that more companies are being funded at the seed stage, so that has translated into a glut of companies seeking more funding at what is known as the Series A venture capital stage (labelled 'A' because it's typically the first professional or institutional funding that a company receives). So take these numbers as a rule of thumb. If you are in the ballpark, you're probably going to have a good chance of attracting investment. If you're nowhere near, you're going to have to hunker down, test more improvements and features, and try to hit the metrics for success.

Chapter 19

Seducing Venture Capital

O nce you've reached the point of product–market fit, you're in
a great place. If you're an A+ performer, your app may already
be making enough revenue to support itself – and even support
expanding your team. The more likely scenario is that you're
going to need to raise money so that you can invest more in engineering and marketing, so that you can get to the point where your
app is generating significant revenues.

So that basically means you're going to be looking for someone
to give you money. There are very few banks with that kind of risk
profile, so your best option is going to be approaching the venture
capitalists we described earlier.

You Need Numbers

The first thing professional investors are going to want to see is
numbers. They want objective measures of great product–market
fit. They want to see a trend – that increasing numbers of people
are downloading your app, that users are spending more time

on your app, and even that they're spending *money* on your app.

If there's a clear positive trend, they know their investment is only going to help accelerate that. While VCs may have a bigger-than-average risk appetite, they are ultimately looking to back a successful business. So don't underestimate how important the numbers are for them.

None of the data you're going to be asked to share should be a surprise. After all, you've been measuring and improving it like a hawk over the last few months.

ENGAGEMENT. The more engaging your app the better. Your conversion rate from acquired users to activated users, and your retention numbers and referral numbers are going to be leading the story here.

REVENUE. You've been tracking this in great detail via your average transaction value (ATV), which you should be able to measure directly. Your annual revenue per user (ARPU) is probably going to be a healthy estimate, and will need to be based on a few months' worth of data (it's unlikely you'll have an entire 12 months yet). Your customer lifetime value will also be an estimate at this point. Be aware if your customer lifetime is already looking as if it is only a few months – that's going to communicate that there is a fundamental issue with your product. You want your potential customer lifetime to be as long as possible. When you look at the core business models – be they gaming, e-commerce, SaaS, enterprise or even consumer audience building – you'll see that all of them have user lifetimes in the years.

CUSTOMER ACQUISITION COST (CAC). Investors are going to want to understand both your user-acquisition strategy and the cost of acquiring each one of those users. With all the channel and campaign experimentation you've been conducting you should have a pretty good idea what this is. With this data, investors can quickly see how much money it's going to take to grow your user base to an interesting size. It will also give them a good indicator of how clever you're being – and how competitive your CAC is compared with other apps in the marketplace.

There is another metric – or, more specifically, analytical tool – that investors will be impressed to see at this point. It's called cohort analysis. Since your app is still pretty young, you're not going to be able to present any long-term metrics. One way around that is to measure – and compare – groups (or cohorts) of users at the same point during their lifecycle with your app.

A typical cohort analysis compares the behaviour of all the users who registered in a given month (let's say that all the users who downloaded your app in January use it on average five minutes per day) with all the users who downloaded it in another month (let's say that customers who downloaded it in February use it on average six minutes per day). By comparing the January cohort with the February cohort, investors can see that users are using the app more – a great sign. Cohort analysis allows you to present a trend – typically from month to month – showing clear improvements across any metric: from session length, to customer acquisition cost, to lifetime value.

Valuation

One thing that you'll want to have in your head is a clear – and justifiable – valuation for your app at this point. While you're shooting for a $10 million valuation, there is naturally a pretty big range at this point. It is going to fluctuate based on numerous factors, e.g. how well developed your app is, how many users it has, whether it's already generating revenue, or even whether it's

profitable already. It also depends on the calibre of the team –
whether you already have the key heavy hitters on board.

The best strategy to get a good feel for a valuation is to do some
research. Check out CrunchBase. It's one of the best (free) collections of information about startups around the world. Find
companies that are similar to what you're doing, or ones that you
would consider comparable in terms of sector, product, team or
traction. CrunchBase records all the dates, amounts and other
details of funding rounds – so you can get a pretty good idea of
valuation ranges.

I've put together a comprehensive selection of links on the
mybilliondollarapp.com so you can find more sources of valuation
and funding information. There are a number of top law firms
who publish this information on a quarterly basis. The diagram[1]
below shows the numbers across all venture-capital-based companies and not just mobile ones, so a bigger and therefore more
meaningful sample. You can see that the average valuation at
Series A funding stage varies between around $6 million and $9
million and has been increasing over time.[2, 3]

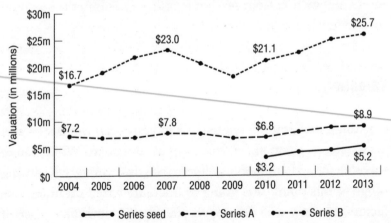

Median Pre-Money Valuation by Funding Round

How Long Does It Take?

You should by now have a good idea about how the venture-capital world works. You should be reading about VCs, meeting them at events and generally getting friendly with them. Typically, they are astute people with good industry and startup experience.

Things will begin to change as you start courting these same people to actually invest in your company. This is where you need to have done your homework – preparing all the numbers outlined above as well as making sure you're targeting the investors who are going to be right for your app.

You're going to have to grow a thick skin. Venture capitalists can be harsh and unforgiving: some will dismiss you outright; others will feign interest with the goal of investing later. One of the first things you should do at this point is figure out which VCs do invest early in Series A funding rounds. A lot of VCs will take meetings and talk with you at any stage to ensure they are building their deal-flow pipeline (and make sure they are building up valuable information about what startups like yours are doing). But be careful, and identify investors with a strong history of investing early so that you don't spend too much time pitching to investors who won't actually be useful at this point.

In the best possible scenario, the process of securing a Series A investment, from the initial meeting to cold hard cash in the bank, is about two months. In reality, it is never that quick. You should allocate between three and six months for your first professional investment round. If you're not well prepared, or your app doesn't have strong enough traction, it may take longer. Experienced entrepreneurs bank on around three months end to end when everything is going well. When things are bumpy with your company, this can easily stretch to six months. This is another reminder why it is so important to generate revenue as early as you can: you don't want to be held hostage to the whimsy of investors. You should be aiming to build a real business as quickly as possible.

From the diagram below, you can see that the average amount of time from the closing of a seed round to a Series A round was 600 days in 2013, or about a year and eight months.[4] The diagram also shows that in about ten years this period almost doubled from just over 300 days in 2004. This is a sign that it takes quite a bit while longer to reach product–market fit. What is not clear from this graph is that the size of seed-round investments also increased significantly in the same period. So, while today startups have a longer runway to achieve results, the quality bar for the end product is also a lot higher.

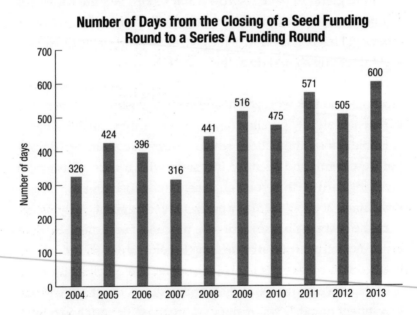

Number of Days from the Closing of a Seed Funding Round to a Series A Funding Round

In the diagram on the opposite page, the good news is that there you immediately notice that the number of seed/angel funding deals has exploded, so there is investment money available. The number of Series A investments has grown, too, but at a much slower pace. So the takeaway here is that there is a high bar to get follow-on investment. You do need product–market fit, you need traction, and you need the numbers to prove it.

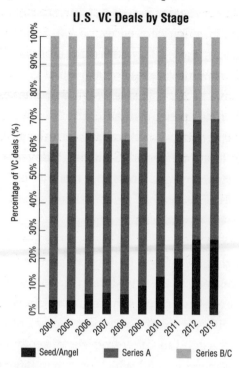

U.S. VC Deals by Stage

Seed/Angel   Series A   Series B/C

VCs in a Nutshell

So what are the key things in mind as you begin a relationship with professional investors? Well, you need to remember that finance is a very lucrative game, and these guys are serious. I think Brad Feld, a very experienced investor and operator, having been on the board of numerous tech companies, including Zynga, with two degrees from MIT, describes our friends in the venture-capital community best: 'VCs only need three rights: Up, Down, and know what the fuck is going on.'[5]

This a simple but clear way to remember what professional investors consider important when they get involved in a deal. But what does it mean?

- Up means *pro rata* rights. When things are going well in a company (going up) a VC wants the right to continue to invest money to maintain their percentage ownership of the company.
- Down refers to the liquidation preference. When things aren't going so well (going down), a VC wants the option to get their money out first, and minimise their losses. This also applies to the scenario when you as a founder want to sell your company, and the VC doesn't have the power to stop you.
- Knowing 'what the fuck is going on' is about sitting on your company's board of directors. While there is some power and responsibility to serving on a company's board, the real advantage here is that a VC has the best possible outsider's perspective about what's actually happening inside a startup.[6]

These thoughts are echoed by Fred Wilson,[7] a VC with Union Square Ventures, one of the investors in Hailo. He points out that these are the must-have clauses that every seasoned VC will want to exercise – with the exception in some cases of demanding a board seat. So it's good to know from the onset which points it is possible to negotiate out and which ones you have no chance of removing.

Big Legals

During the seed-financing round, working with standardised documents was a great option. In the Series A round, now that you're dealing with big-time investors, that's no longer going to be possible. Now that the financing requires bigger numbers – with bigger options for both upside and downside – investors are going to demand additional rights. Hence the agreements are going to be more complicated, and specific to your company and situation. That means good news for lawyers, and expensive news for you.

You're going to have to get ready to spend between £20,000 and £40,000 on legal fees (that includes the investor legal fees, which you'll be expected to cover).

Before you get too serious, make sure that you've established a relationship with two or three legal firms. They will spend time with you (usually for free), taking you through the process. Make sure that you get a good introduction from a friend, or, if that is not an option, visit the Billion-Dollar App website, where I've put together a list of some of the top firms in a number of capital cities.

Now, let's have a look at what the process looks like.

Term Sheet

Similar to the seed-funding stage, the first part of putting a Series A deal together is to agree on the major deal terms, which are summarised in a term sheet from the investor. It's good to try to focus on getting the more flexible investor(s) to produce the term sheet, as it will allow you to bring in other investors on similar terms.

In this round, the core elements are going to change a bit. The major term-sheet elements are the following:

- **PRO-RATA RIGHTS.** This allows an investor to continue to invest money to maintain their percentage ownership.
- **LIQUIDATION PREFERENCE.** This allows an investor to get their money out first in the case of an exit or other scenario.
- **PREFERENTIAL RETURN.** Investors will try to negotiate at least a 1x, if not a 2x, return on their original investment before other classes of shares are paid out. Be careful about this provision: 1x is normal; 2x is bad.
- **BOARD SEATS.** Your lead investor will insist on a board seat, especially if they are investing a considerable amount of money. Remember to select your board members very carefully, and make

sure that you can maintain majority control of your board. Investors often want observer rights either alongside a board seat (or in lieu of) as well as asking for information rights (guaranteed updates about the company's performance on an agreed frequency).

- **PREFERRED SHARES.** Investors will require that you create a new class of share that has voting (and other) rights above and beyond that of common stock.

- **EMPLOYEE OPTION POOL.** You will need to formalise an employee option pool (typically 10–20 per cent of the company) to ensure that you have a mechanism by which to attract and retain top talent.

- **EMPLOYEE AGREEMENTS.** If you've set things up properly these will be in place. Investors will want to see that there are adequate intellectual property assignment provisions, confidentiality provisions and non-competition/non-solicitation provisions in place.

- **FOUNDER VESTING.** Most investors will try to add a guarantee to ensure the founders stay around for at least four years by making their founder shares vest over a fixed schedule. Most investors are adamant about this; some are more flexible. Recall the benefits that I outlined about adding this in yourself, especially if your company has several cofounders, when we discussed the million-dollar app.

These are the main elements designed to give you a flavour for what you will be negotiating. It's by no means meant to be a comprehensive list. Due to the involved nature of startup legals, I'd suggest going through the resources on the Billion-Dollar App website when you want to dig into it more.

The Clauses that Can Screw You

Having worked with a lot of investors over my last few startups, and having completed an exit as well, I can attest at first hand to the importance of understanding all the major legal and contractual details that will directly affect you, the control you have of your company and how much you will earn in the event of an exit.

Here is a list of the most contentious clauses, what they mean and what to watch out for. It's all quite legal – but you'll be a happy person if you get these negotiated in your favour.

- **LIQUIDATION PREFERENCE.** This is one of those clauses you don't need to get stung by. It's standard for investors to get a 1x liquidation preference.

- **ANTIDILUTION RIGHTS.** Investors will ask for this in the Series A funding round. Basically it means that an investor has the right to maintain the same percentage ownership of your company automatically if the value of the company goes down. This is really tough on the founders. It means losing more of your company's equity *automatically* in a down period. It's unlikely that you will be able to remove this provision – but you need to be aware of it.

- **NON-PARTICIPATING LIQUIDATION PREFERENCE.** Liquidation preference determines how the pie is shared in a liquidation (or exit) event. A founder should always seek to have a non-participating liquidation preference for investors. Without going into too much detail, a 'participating' liquidation preference allows an investor to double-dip in terms of the slices of pie that they are entitled to. You'll find a more detailed explanation on mybilliondollarapp.com.

- **DRAG ALONG RIGHTS.** This provision grants the investors the right to compel the founders (and other shareholders) to agree the sale, merger or liquidation of the company or block a sale, merger or liquidation. As a founder you do not want to be dragged along in this case. Ideally you should be able to negotiate that investors would only have the right to block the sale if the return was below a certain level.

- **WARRANTIES.** In investment agreements in Europe it is very common for founders to personally give basic assurances that everything the company is doing is proper and correct, these are known as warranties. In other countries, like the USA, founders don't give warranties at all.

Again, do check out mybilliondollarapp.com for more details, because this is definitely a complex area. That said, there are always going to be numerous details that are specific to your company and situation. The best possible advice here: make good friends with a few lawyers so that you always have someone to call on with a specific question.

Employee Legals

Another area where professional investors are going to get involved is employment contracts. Naturally, this starts with the cofounders. Since the investors are giving you a huge chunk of money in exchange for a part of your company, they expect you to hang around for a few years, so the first big thing, which we saw above, is going to be the addition of a founder-vesting schedule.

What else are they going to insist on adding?

One thing that I have seen investors try to do is remove any accelerated option-vesting scheduling on exit or any change in control. This means that for either the founders or other early employees there will be no right to accelerate the vesting of their options in the case that the company is sold (or in the case of IPOs, but that takes many more years). The reason investors want to remove this is to ensure that people stay with the company when it's acquired. It's one of the main reasons companies are acquired: for their talent. So giving key people a way 'out' reduces the value of an acquisition. What is reasonable, and I referred to it earlier, is to give key employees a 'double-trigger' vesting. This means that two events need to happen to trigger accelerated vesting, the more important being that an employee adopts a role less important or senior in the case of the change in control.

One other very foreign concept to 'normal' business, but very common to venture-capital-backed businesses, is that of 'good leavers' and 'bad leavers'. Basically, these are clauses inserted by investors (or the company's board) as an insurance policy. If the board deems someone a bad leaver, that usually means that person can be stripped of all their benefits – specifically all their options – whether they are vested or not. So that's a pretty big deal – and underscores yet again why you want to make sure you keep control of your board of directors.

Investors will want to ensure that all the current (and future) employees sign new, revised employment contracts, should your original ones be a bit too lenient. In addition to the above clauses, it's standard to include non-compete clauses (so that employees can't create a new company that directly competes with the business, or go and work with a direct competitor), anti-poaching clauses (so that, if you leave, you are forbidden from trying to take other members of the company with you) and a variety of clauses ensuring that all the intellectual property developed while you were working for the company is owned by the company, rather than the employee.

I hope I have painted a clear picture of what it means to take money from professional investors. The process is definitely not for the faint of heart. It is a long and involved process. If you're truly serious about building a great technology company – at breakneck speed – then this is definitely one of the best strategies to pursue. Having great venture-capital investors on your side can accelerate the process immensely, and give you the resources to learn and execute at a phenomenal rate.

Summary

Achieving product–market fit is probably the most exhausting – and tricky – part of building your company. This is the time when real entrepreneurs shine. This period is also particularly challenging, because you also need to build the foundation of a real company at the same time.

Relentless focus on understanding your users – and focusing on the problem you're solving for them – is the key to success. If you settle for a weak level of product–market fit, you are constantly going to be struggling as your company moves forward. Your product won't be in demand, you won't be competitive in the face of other apps, you'll struggle to persuade great people to join the team and you'll struggle to generate the revenues and lifetime value you need to generate big margins. In short, you won't have built a good enough product.

And there's an even bigger risk that results from a weak product–market fit: your vision for the future will be cloudy. If you can't deliver something that your users *really* love *now*, then what sort of vision is driving the future of your app? Are you going to keep building OK features for your users? Are you aiming for being a decent app – one that users 'like' but don't quite 'love'? That is not going to get you into the Billion-Dollar Club.

So, if you're not quite there yet, don't succumb to trying to grow too quickly without the right product–market fit in place. Success is understanding what people *really* want – and translating that into an app they are willing to pay money for. Once you master that process, you will be in the powerful position of being able to repeat it to develop new features and new products – to make your users love you *more*.

Without product–market fit in place, trying to grow your user base and fine-tune your business model is going to be more complicated. People need to love what you're offering before you can optimise how much you're going to charge them for it.

On the flipside, once you see users clamouring to use your app you'll be in a great position to start building a robust business. You'll have the chance to keep wowing users, introduce more features, figure out how to make big revenues and then get on the growth path to hundred-million-dollar land.

STEP 3

The Hundred-Million-Dollar App

Tuning Your Revenue Engine, Growing Users and Raising Series B Funding

Going From a Ten-Million-Dollar App to a Hundred-Million-Dollar App

App

- You've nailed product–market fit and have an app that users love. Now you need to keep – and grow – that traction by systematically improving every part of your app and focusing on how you're going to generate revenue. As growth continues, you'll need to think about international expansion and the complexities that will mean for your product.

Team

- You have a great core team in place and you're still operating in a lean fashion. You're now ready to fill in a couple of key leadership roles – the first focus will be in marketing and/or sales to drive your growth engine.

- Figuring out your ideal organisation and team structure is also going to be critical as you prepare yourself to scale up.

Users

- You have a solid user-acquisition strategy in place – with a very good idea of customer acquisition cost – and a decent idea of annual revenue per user. Luckily, you now have the budget to start executing that strategy.

- As you build your marketing team to drive your growth engine, you'll also need to start getting in a user-retention and -referral strategy.

Business model

- If you have an e-commerce, marketplace or gaming business model, you have already validated that users will pay for your app; it's now a race to improve lifetime value by holding on to users, and then improve margins to make sure your company has a path to profitability.

- Your goal will be to get your business model humming – which means that, in addition to making users happier, you'll need to optimise price points and drive down customer acquisition costs.

- You may also discover that you need new features – and revenue streams – to boost your model and chase higher profits.

Valuation

- In the last section $10 million was the average valuation for all Series A software startups (including apps).[1] To hit $100 million at this stage you're going to have to pull out all the stops. While the average Series B valuation is around $27 million, this is the point when truly great companies pull away from the pack. They don't double their valuation: they blow it out of the water.

Investment

- You've landed your first professional investors – and are finally playing with the big boys. You now have a couple of million dollars in the bank (or more, depending on your traction and negotiation skills).

- While this money yields some breathing room, it's not a huge amount in the scheme of things. Hence your focus has to be on delivering on a profit-generating business model.

- If you deliver on your revenue engine and start driving your growth engine, you will have a massive advantage when raising more money. If you hit profitability, even better – then you can take investments entirely on your own terms.

Chapter 20

A Colorful Lesson

'What if you threw a $41 million party and no one turned up?'

#BILLIONDOLLARAPP

Nailing a Series A investment is no small feat. There are even a few companies that manage to raise a *couple* of rounds of financing before they launch to the public. Occasionally, visionary entrepreneurs create a product so different and so game-changing that it whips up a frenzy.

This is precisely what happened towards the end of 2010 through to the beginning of 2011. A couple of serial entrepreneurs, Bill Nguyen and Peter Pham, created an app that had even legendary Silicon Valley investors Sequoia foaming at the mouth.

The company landed a massive $41 million in funding. And that was *before* it had released anything to the public. To say the anticipation was electric was the understatement of the decade.

That app was called Color.

A Better Facebook

I love entrepreneurs who want to change the world. 'I thought we were going to build a better Facebook,' says Nguyen, Color's CEO.

Color's launch was amazing. People started downloading the

app so quickly that it rocketed, briefly, to number two of the most-downloaded social-networking apps – just behind Facebook. So he got that bit right – but then the problems started.

The first night after the launch, everything came crashing down. They started noticing that users were rating the app at one or two stars out of five. That was the first serious warning sign.

Color was a photo-taking and -sharing app that allowed users to see photos from nearby events, creating 'elastic' clusters of people using the app, grouped together by their proximity. The idea was clever, and wasn't very clearly explained in the app.

The problem the app suffered was actually quite simple: in order for Color to work, lots of users had to be in a similar location, but, since the app wasn't seeded across a big population of users before the launch, users would arrive at the social network and see no one close by. It was a ghost town. I think the *New York Times* said it best: 'What if you threw a $41 million party and no one turned up?'[1]

'Within 30 minutes I realised, Oh my God, it's broken. Holy shit, we totally fucked up,' says Nguyen. 'I thought we were going to build a better Facebook. My reaction was like putting your finger into a light socket. You know something went very wrong.'[2]

A $41-million lesson

The point here is that no matter how good a salesperson you are – and no matter how impressive your experience is – you can't escape real user feedback. You can't escape reality, just as you can't fake product–market fit. Be brutally honest about measuring product–market fit. If you're not there, go back and sort it out.

If you're a great salesperson, you will find the money – but ultimately your users will determine whether your app succeeds or fails. It seems like Bill Nguyen had enough cash in the bank from previous jobs, so he'll be fine.

Chapter 21

Tuning and Humming

At this point in your journey, revenue becomes critical. It's not just about getting dollars in the door, though: to make it to the top of the app heap you need to have revenues that scale. Steve Blank, a serial Silicon Valley entrepreneur and the father of the lean-startup movement (we mentioned this in Chapter 7: 'Getting Lean and Mean'), explains it brilliantly in *The Startup Owner's Manual*: 'Simply put, does adding $1 in sales and marketing resource generate $2+ of revenue?'

Your goal is to create a very efficient revenue engine – and keep it evolving. You want to get more out of it than you're putting in. In this part of the book we'll investigate how to tune and adjust your revenue engine to make it deliver magic.

By now you're generating healthy revenues (otherwise, people wouldn't have invested in you and valued your business at $10 million). Alternatively, if you're focusing on a consumer-audience app shooting for an advertising-based business model, you will have built an app so amazing that people are downloading by the millions, and using it addictively every single day. You will also have a small but robust team in place (probably 10–20 people).

You've also nailed your first round of funding from professional investors (your Series A). You have now turned what was just an 'app' into a real business. This means that you're going to have to ignite your growth engine and refine your revenue engine so that, by the end of this stage, your app is generating millions in revenue. Most importantly, you're going to need to put in place a robust plan for your organisation (an 'org chart') and identify the key people you need to hire to make your operation run smoothly, and set the foundation for even bigger growth in the near future.

Your goal is to get all the aspects of your business humming. When you master that – and it could take anywhere up to a couple of years – you will have the choice of how next to develop your business. Since you'll be generating revenue (and will ideally be profitable) that means you won't (necessarily) have to rely on investors, because you'll be the master of your own destiny. And that also means, should you want to grow faster than your means allow, you will be in a significantly more powerful position to negotiate with investors.

By the end of this section you will see your team increase in size substantially. Depending on your business model and revenues, it may reach anywhere from 25 to 75. Hailo ended up with a team of about 100 people by December 2012, when it closed its Series B. Naturally, team size is highly variable. It is dependent on everything from the calibre and experience of your employees, to the complexity of your product, to the market you are operating in. You may not need a big team yet but, once you're generating significant money, the option becomes available to you.

At this stage you're going to have to start driving the core parts of your business simultaneously – that means focusing on business model, profit margin and product development. As these components develop, more paths will open up for you. There isn't going to be an ideal team size, revenue target, profit margin or funding goal. We'll go through numerous paths to get to the $100 million stage – and it certainly is going to be exciting.

Where Are We Going?

At the same time as making plans for world domination, you also need to make sure that your business is self-sustainable, that the app is generating enough revenue to pay for salaries, office space and all operating expenses, while also delivering attractive profits (or at least showing clear promise). This is what ultimately every business is looking to achieve – even startups.

To get to that point we're going to be focusing on three things in this section.

1. TUNING YOUR REVENUE ENGINE. You need to make sure that your business model really scales. Are you going to be profitable with a few thousand users? Or do you need to attract (and retain) *millions* of users? Will your same business model appeal to people all over the world? Will everyone pay $0.99 to download it? Or will everyone be equally willing to make in-app purchases? Will some users be a lot more valuable than others? If so, should you spend a lot more acquiring more valuable users? All this needs to be figured out.

2. IGNITING YOUR GROWTH ENGINE. Growth means getting your app into the hands of lots of people – start thinking millions and then think globally. Is your current user-acquisition strategy going to scale? How are you going to tweak your app to hook millions of users? Will it need to be customised by region or country or language?

3. YOUR TEAM. You need to get the right people in place to make this work. 'People are your greatest asset' is a mantra because it's true. In the world of technology and software it's *more* true. Your app software is only as good as the developers, product managers, designers and testers who are putting it together. Your ability to attract users and generate revenues is only as good as your marketing team. And your ability to make informed, timely decisions is only as good as the data and analytics engineers you have on board. You need to think about how you're going to build that team and, more importantly, put a plan in place to retain them.

Once you have these elements under control, you'll have a true mobile-app company. At that point you'll have a very rosy future – and a bit of breathing room to think about how you want to proceed. This section of the book will walk you through the steps you need to take, and tell you how to navigate through them effectively. If you still have your vision set on becoming a member of the Billion-Dollar Club, you'll have the best possible foundation from which to scale your business.

At the end of this chapter – as you approach $100-million-dollar land – you'll be in a position to go after Series B funding. Securing this funding is what will enable you to explode your business – to go from being a competitive player in a market to becoming the undisputed leader. That is a lofty goal, and, as you can imagine, there are a lot of pieces of the puzzle that must be put in place before this is feasible.

As you get close to this stage you'll face some big strategic choices. You'll be faced with questions such as:

- Should I go for growth (acquire more users, generate more revenue, at the cost of profitability)?
- Should I aim for profitability (sacrifice some user and revenue growth but retain more control and potentially position myself to fund my own growth and not give away more of my company to investors)?

It's worth spending time mapping out the vision – and the mission – for your business. This will help shape your future direction. Naturally, as your business grows, there are going to be a lot more varying points of view – all of which you need to manage and take into account. There is no single correct way but there are certainly lots of proven tactics to adopt, lots of potholes to avoid and many things you can do to mitigate unnecessary risk and put yourself in the best position to reach that elusive billion-dollar success.

Portfolio Brothers and Sisters

As a result of your successful Series A funding, your company has changed. Not only do you have a new board member (or two) with the arrival of your first venture-capital investor (hopefully, you haven't given away too many board seats at this stage), but you have also graduated to the level of a proper 'portfolio company'. What does that mean?

By accepting their investment, you have become part of the investor's family of businesses. Top VC firms such as USV (Union Square Ventures), Accel Partners, Atomico, Index Ventures and KPCB (Kleiner Perkins Caulfield Byers) encourage lots of interaction among their portfolio companies, to share information and advice, to hang out together. The better the VC, the more widespread and active and well organised this 'social' programme is. Being a member of an elite group of companies brings a slew of benefits, namely access to the experience of many other entrepreneurs, often very experienced ones.

What is also worth noting is that VCs tend not to invest in competing companies – so it's unlikely (though by no means impossible) that you will have a direct competitor as a portfolio company. Wise investors treat this as a golden rule – since they want to be seen to be backing their 'preferred' horse, rather than betting on multiple winners in the same race.

In addition to the standard CEO forums hosted by these firms, more innovative investors such as USV bring together the product, engineering and marketing teams from their portfolio companies a few times a year. This allows a free exchange of information, challenges and, most importantly, solutions. Friendships also form very quickly and easily in these types of environments – and provide a great support network as you grow your business.

Being an entrepreneur is a notoriously lonely job, involving long hours doing anything and everything to transform your vision into reality and often having to battle failure after failure before reaching success. And, on top of that, the only people who can understand your plight are other entrepreneurs, who are suffering the same stresses and demands as you. So this additional angle provided by venture-capital investors can be a huge boost. My advice is: make the most of it.

Finding a Marketing Team

With your Series A funding what you need to invest in is a leader for economic and growth engines. The person you need to hand the reins to is a top VP of marketing.

Attracting users to your app is damn hard work (there are some great exceptions, but even they benefited from having a superb individual manage things behind the scenes). The person in charge of this is your VP of marketing. They are not only in charge of acquiring users, but are charged with retaining them. Today, if you want to be successful, you must demand that your VP of marketing be also a big-time data geek.

Managing all that complexity – and the team that brings it all together – is no small job. And being able to grow your user base means getting an amazing VP of marketing on board. It will take the load off the founders and CEO, and put it in the hands of a pro. They'll need to build up a team and deliver a robust plan to deliver reliable growth.

How to hire a VP of marketing

I have been lucky to have hired and worked with two truly amazing marketing guys. The first was Steven Sesar, who joined the team at WooMe having previously led the marketing for Shopzilla (which was acquired for a cool $560 million).[1] He was a numbers guy through and through – and a huge part of our success was his ability to cost-efficiently acquire and retain users.

Carl Lyons is, at the time of writing, the VP of marketing at Hailo. Before joining Hailo he had led a variety of great digital-marketing teams – namely at Lastminute.com, Capital Radio and the *Guardian* – but is equally a brand man. Brand is increasingly important in a noisy, highly competitive app ecosystem.

So what's the checklist that you should be focused on when bringing someone in to head up your marketing efforts?

DATA AND ANALYTICS. Recent figures show that 48 per cent of app marketers' greatest mobile-advertising concern is ad tracking and measurement,[2] so what you need first and foremost is someone who gets data, who gets analytics and who gets conversion. Without those skills you're pretty much dead in the water. Make sure this person has cut their teeth at other top tech companies running and managing big app and online marketing teams.

BRANDING. A great marketer knows all about brand. That means presenting a single consistent and powerful profile of your app to everyone in the world. This encapsulates everything from the name, logo, visual design and advertising to the tone of voice and the copy used on your website. If you have a great VP of product, then this person and your VP of marketing will work like two peas in a pod to deliver on this vision.

INTERNATIONAL EXPERIENCE. The app world is international. If your VP of marketing doesn't have international experience then they're not going to be very useful. International audiences in the app and online worlds behave very differently and in an ideal world your marketing head will have had great exposure to this. When money is tight, you don't want them to be learning on your dime – they should already have joined you with that experience.

TEAM BUILDER. You need this person to know how to build and energise a team. At this stage in your business, your marketing organisation will need to scale, and you need someone in place who has experience of doing this effectively. This means not only the ability to hire great full-time employees, but also an understanding of why and when to hire an agency or freelancers as opposed to full-time team members.

AGENCY EXPERIENCE. This is a mixed bag. There are many companies that just don't use one – and, frankly, agencies are becoming less relevant to startups and more of a crutch for big corporates. Unless they are super-specialised in terms of mobile-ad media buying and optimisation – e.g. players such as Fiksu – then agencies are not much use beyond a bit of creative or PR work. And, anyway, having a good marketer internally usually solves the advertising creative component in the early days. Outdoor advertising for most mobile players doesn't drive app downloads or conversions, but is good for branding and awareness (if you have the money to burn) or if your business requires it (Hailo used outdoor to communicate more with taxi drivers than passengers to demonstrate its commitment to building a genuine business).

Outside the skill set above, you need to make sure your VP of marketing is in it for the long haul and truly loves the brand, the company and your vision. This person will be instrumental in communicating your vision for your app to the rest of the world.

Negotiating terms

In terms of remuneration, there are a couple of rules of thumb. If the VP of marketing is coming in at a pretty early stage – is one of the first 10 or so people, say – then you're going to be talking percentages (e.g. 1 per cent, 2 per cent, 5 per cent) of your company's equity. As the team grows, you're typically going to give employees options as a percentage of their base salary. In the case where the VP of marketing comes in a bit later, and is getting paid £150,000, they would typically have a multiplier of 0.5^3 – or £75,000 of options at the time they join. So, if the company increases in value 5x, then they will have £375,000 worth of shares, or, in a great case where the company value increases 10x, they will leave with a handsome £750,000.

Getting hands-on

Now that you have someone in place who has done it all before, it's a good moment to breathe a sigh of relief. If you've hired well, your VP of marketing will be already thinking about whom they want in their team and will already be warming up contacts they have from previous jobs.

Tuning your growth and revenue engines is going to be a good amount of hands-on work: chasing user-acquisition channels, testing campaigns, negotiating partnerships, working with PR channels, digging through data, setting key metrics and goals.

Ideally, your marketing team will have another experienced and sharp marketing manager on board with deep digital experience, and hopefully a younger (and cheaper) analyst to help with the crunching of all the numbers. You don't need a big team in place right now, but you need one that is going to keep experimenting, keep iterating and delivering a solid path to profitable user growth.

Chapter 22

Getting Shedloads of Users

L et's start this chapter with your goal, because it needs to be a bit more finessed. What you need to be able to do is hit your user numbers, so that they deliver a revenue number that comes with a desired profit at the end of every quarter, and then at the end of the year.

You need to be able to model how to effectively get shedloads of users who will translate into a profitable revenue stream within a certain amount of time, say the next 12 months.

You should be thinking about how to model this in two ways:

- Model your user, revenue and profit growth based on your current funding and cash flow
- Model your user, revenue and profit growth with an additional round of funding.

These two scenarios are essentially what you'll be discussing with investors – once you've demonstrated that you can reliably and efficiently add users and generate revenue.

One of the most frequent mistakes I have heard about is using too few traffic sources to acquire new app users. The best mobile marketers I have met are constantly trying new channels, sources and networks. It costs relatively little to try a new network, and it can make all the difference. This is especially true when you're expanding internationally and are facing very new and different channels – and very new and different users. Given how quickly advertising sources can change, identifying overperforming mobile sources and taking advantage of them can yield fantastic results. It's also important to keep in mind not saturating any particular channel at the risk of annoying a given audience.

At Hailo, we found it possible to improve our mobile advertising by about 3x to 5x by focusing on a larger number of better-performing channels. In addition to just the normal paid advertising networks, we very successfully worked with Facebook and its mobile-advertising platform (it worked great in every market except London, funnily enough). Twitter mobile advertising reaped some pretty good results as well.

Traditional Channels

A mistake many app businesses make is to rely on traditional marketing channels. Traditional marketing channels are too general. Channels like PR, TV or radio are not only expensive but they don't have the ability to be laser-focused. Not only that, but traditional marketing channels have been proven not to move the needle on app downloads. Even traditional online advertising – with a desktop focus – doesn't convert well on mobile at all. So why bother doing it? You're just going to be wasting money.

I'm not saying that you should shun traditional marketing entirely. It's important to have some amount of desktop web advertising to help drive activity via social media, Google, etc. but don't expect those channels to be particularly efficient download drivers. They will, naturally, be good traffic drivers to your website or social-media profiles.

Mobile User-Acquisition Channels

In the United States 91 per cent of all Americans have their mobile devices within reach 24 hours a day, seven days a week,[1] so it's a bit of a no-brainer that mobile advertising is also growing at a crazy rate in an attempt to capture and direct people's attention.

Fiksu means smart in Finnish. It's also the name of a very cool company founded by Micah Adler, a Finnish-born, US-raised entrepreneur with a degree in mathematics from MIT and a PhD in computer science from the University of California in Berkeley. Before he started Fiksu he was actually a professor at the University of Massachusetts.

Adler was unhappy about the early waves of mobile advertising: 'We started to develop algorithms to make digital marketing efficient. It was costing us $3 to generate a download of our app, which was too expensive.' Employing his rather large brain and substantial academic qualifications, Adler and his team were able to massively optimise the mobile-advertising process. 'Over the next six months, we went from a cost per download of $3 with thousands of downloads per day to $0.26 per download with tens of thousands daily,' he explains.

The core of what Fiksu does is simple: its system makes sure that, first, it's targeting the right users for your ad and, second, it finds as many people as possible who are going to be interested – targeting and reach. It deploys its fancy algorithm over more than 200 mobile-ad networks (that's about 99 per cent of the available mobile-ad inventory out there) and does all the calculations in real time.

For example, say that Fiksu is working with one of its mobile-advertising network partners, and a user's Android phone gets turned on. The advertising network signals to Fiksu that there is a device available ready to serve an ad to, passing on details such as the ad format, size, location of the user and other details to better target the ad to the user.

At this point Fiksu has 100 milliseconds to decide whether it wants to buy this ad space (pretty dynamic, eh?) and, if so, which ad and at what price it should pay for it. The Fiksu platform makes these kinds of decisions 50,000 times a second.

Pretty cool, right? Fiksu works with companies big and small. It can improve your mobile-advertising effectiveness by a factor of three, either by increasing the number of app downloads three times or by cutting the cost to generate the number of downloads you want by two-thirds. At Hailo we used them extensively and they delivered the goods.[2]

Incentive-Based Networks

Incent networks, as they're also called, are a relatively new entrant in the world of user acquisition. It seems that, as long as mobile developers search for new ways to drive downloads, entrepreneurs will keep dreaming up new business models to give them what they want.

The basic premise is that a network will effectively give vouchers (from a variety of big partners such as iTunes, Amazon, Groupon, Hulu and Google Play) to users in exchange for their downloading and trying other apps. Interestingly, one of the biggest (if not the biggest) players in this market is FreeMyApps.com – funnily enough created by Fiksu.

FreeMyApps touts itself as an 'app discovery network', with a community of 1.6 million highly active users dedicated to discovering primarily new games. The site also has more than a quarter of a million followers on its Facebook and Twitter accounts – which you can also target. So do check into the profiles of users who use these types of sites.

After talking to a number of marketers about these channels, you realise that the feedback is quite mixed. Often the channels are helpful in bumping up your download numbers, but they don't typically translate into valuable users. I wouldn't rely on it as a primary user-acquisition channel – it will probably just lead to a lot of zombie downloads.

These types of networks seem to be springing up all over the place, so it's well worth doing your research, because they may work for you. If nothing else, it is another channel to experiment with.

Mobile Social-Media Channels

The rise of mobile is pretty obvious (I've been harping on about it for a few chapters now). What's interesting, however, is that social media are also now driving real e-commerce. Mobile now accounts for 40 per cent of time spent on social media. Facebook has passed the 50 per cent mobile-usage mark and Pinterest is at 48 per cent. Together, these two players are responsible for 56 per cent of social-media-generated e-commerce.[3]

Social media still represent a small source of direct e-commerce traffic but they are expected to drive $30 billion in e-commerce sales globally by 2015 – $14 billion in the US alone, according to one study.[4] We know that social media do play a very important role in multi-touch attribution, as 74 per cent of consumers rely on social networks to guide their purchases, according to research firm Gartner.

The graphic[5] below is testament to how powerful *mobile* social-media channels are in influencing purchasing decisions. Twitter users make over 50 per cent of their social-media purchases on mobile. Pinterest and Facebook are not far behind, with the mobile channel responsible for about a third of their purchases.

The rise in purchasing via mobile advertising means that a massive increase in marketing and advertising spend is happening on social networks. There is definitely a rise of social-led commerce that can be further flamed by paid-for advertising on social networks.

In 2012 global mobile-ad spend was up 83 per cent on the previous year, to $8.6 billion.[6] By 2015, mobile advertising is projected to grow to $33.1 billion, which will be 25 per cent of Internet advertising and 6 per cent of all advertising spend.[7]

Social media advertising will balloon into an $11 billion market in 2017, about 20 per cent of which will be generated by mobile.[8]

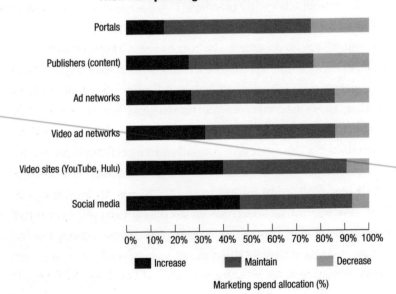

How Marketers and Agencies Plan to Allocate Their Ad Spending in the Next 12 Months

Marketing spend allocation (%)

In terms of the channels that are getting the most attention from advertisers, the chart opposite shows that social media are clearly leading the charge. As social becomes an even larger revenue driver, more advertisers are going to pile into the channel (and drive up costs). Similarly, video ads, in addition to video sites such as YouTube, are demonstrating that they can drive big results as well.

Facebook

In my experience, one of the most effective mobile-marketing channels is Facebook. It has worked like a charm for Hailo for a number of reasons: before launching a marketing campaign using Facebook, Hailo was already aware of which demographic would benefit from a more efficient way of hailing cabs. Using the profile information available on Facebook (city, age, gender) Hailo was able to easily target users who fitted its demographic. It was also able to target users who were most likely to be interested in Hailo, as they had 'liked' similar products to Hailo, such as other taxi or black-car services.

In marketing your app though Facebook, note that Facebook Mobile is an obvious channel to pursue, as the majority of users access Facebook from a mobile device. In the US, 78 per cent of Facebook users access the social network via mobile – that's 100 million people.[9] That number will hit 154.7 million[10] by 2017. So rest assured that this channel has legs for some time to come.

Facebook also allows you to buy native in-stream ads. Facebook ads are hugely effective: a recent report published by Nanigans, one of the largest buyers of Facebook ads, revealed a 375 per cent explosion in click-through rates during 2013, and a whopping 152 per cent return on investment.[11]

Another interesting thing to note is that Facebook Mobile ads are 1,790 per cent more profitable on iOS than on Android.[12] Nanigans surveyed more than 200 billion ads and found that return on investment for Facebook Mobile ads is a shocking 1,790 per cent higher on iPhone than on Android. On top of that, revenue per click was 6.1 times higher on iOS than on Android for the first three quarters of 2013.

This is just meant to be a flavour of what you can do. Check out mybilliondollarapp.com for more details.

Chapter 23

Revenue-Engine Mechanics

You and your marketing team have invested a huge amount of time experimenting with user-acquisition channels and campaigns. You now have a deep understanding of the customer lifecycle stages in your app and you've been optimising your conversion metrics. And, now you're generating real revenues, it all feels good.

The focus now is a tough one: it's to find a scalable revenue model. It's actually a bit more complex than that. You want a scalable *economic* model. Success means managing not just the revenue component but also the costs. In a hugely simplistic nutshell, you want your customer acquisition cost (CAC) to be less than your lifetime value (LTV). Not brilliantly insightful, I know, but it needed to be mentioned. Don't worry: we'll dig into meaningful ratios and not just commonsense observations.

To get to the point of billion-dollar success, however, you need to nail both of these – and there are only two ways to do it. Either you focus like a maniac on driving customer lifetime value through the roof (once you get a user, you need to keep them spending and spending), or, you figure out a way to attract millions of users to your app while simultaneously driving your customer acquisition cost as close to zero as you can.

It takes a genius combination of idea, market, product, growth engine and revenue engine to achieve this. Let's dive into both strategies in a lot more detail.

Exploding User Value

Some apps are just built on the back of ever-increasing lifetime value. This works for companies focusing on e-commerce and marketplaces. Why? If they succeed in delivering a product or service users love, well, users just keep coming back and buying it. Amazon and eBay are traditional masters of this approach. They focus on utilities, on necessary products and services, and frequent purchases. And then they make the experience so good – great website, great app, great selection, great delivery – that you just keep coming back again and again.

Uber, Square and Hailo are the new mobile utility equivalents. They generate money from Day One, and seek to build huge audiences and deliver an amazing experience. They focus on common, high-price transactions. Their success is based on keeping users. The better the product the longer the customer lifetime. That's great upside.

Another category of app business model falls into the category of lifetime value maximisation. Software-as-a-Service (SaaS) apps and enterprise ones need to attract and maintain large user bases to generate large subscription revenue streams. Companies like Dropbox and Box have excelled in this area, with their app channels representing big parts of that revenue (and big factors driving user retention as people sync data and photos to the cloud).

In the SaaS case there are two widely shared rules of thumb. First, your lifetime value should be at least three times your customer acquisition cost: if a customer typically spends $100 over the course of their lifetime with you, you shouldn't be spending more than $33 to acquire them. That number makes sense because SaaS apps tend to have higher user-acquisition costs, and also have a tendency to churn (i.e. customers leave and stop using the service). If you look at bigger SaaS and enterprise players you tend to see lifetime value grow to five times the customer acquisition cost.[1] This shouldn't be surprising, since classic enterprise software companies are geniuses at locking customers into their platforms.

There's yet another quite important consideration in this equation. You want to be thinking not only about how long it takes customers to start generating revenue for you, but also about how long it takes that customer to become profitable.

This has a big impact on your bank balance. Let's think about two scenarios. In the first case, imagine that a customer has a lifetime value of $100, and they cost $25 to acquire. Nice – a 4x multiple is very healthy. Unfortunately, imagine that customer spends that $100 over the course of five years – the equivalent of $20 per year. Hmm. That means you don't break even until Month 15 – i.e. it takes 15 months to recoup the cost of acquiring that user, and they are profitable only after that point.

In the second case, we have another customer with the same lifetime value of $100, and they also cost $25 to acquire. But for whatever reason they spend their $100 within 12 months and then leave. That means they are at break-even after only three months. That means you start earning a profit very quickly – and, most importantly, you can reinvest the profits to acquire new users.

So the time it takes to earn back that acquisition cost is important: the longer it takes to earn back, the more money you're going to need to raise.

Zero User-Acquisition Cost

The other option is to figure out how to attract a stupendous number of users to your app at virtually no cost (i.e. drive that damn customer acquisition cost to zero). With a huge and growing audience, no matter what you do you will (probably) make money. One valid strategy is to focus religiously on engagement – and retention – and then wave your stats under the noses of companies with ready-to-plug-in monetisation engines (specifically advertising platforms), such as Google, Facebook and Yahoo!.

Our friends Waze, Snapchat and Instagram all went down this route. It's not the most reliable strategy, but, if your app happens to be in the right spot when that swell of users turns into a once-in-a-lifetime point break, then you need be ready to ride it all the way. These apps have achieved viral growth – all at different rates – and have achieved billion-dollar valuations (or sales) due to the massive power of free user acquisition.

While that approach is a bit too much of a perfect storm for me, Flipboard exemplifies a more healthy approach to audience building, while also developing a robust revenue model. Not only has it developed a user base in excess of 100 million, combined with robust retention, but it has also developed its own revenue-generating advertising platform. Its innovation has been swift, too: in addition to advertising, it has launched affiliate-type, embedded e-commerce, transforming magazines into stores as well.

Gaming Apps

I love playing gaming apps, and I have to admit to a secret affair with Candy Crush Saga. I am one of the people who have played it 151 *billion* times during 2013.[2] I am one of the 200 million people who play it every day.[3] But I am *not* one of the people paying to play by making in-app purchases – that is a tiny sliver of its users. The app generates anywhere from $1 million to $3 million per day, which translated into about $1 billion in revenue in 2013. Let's dig into that number, though. Let's assume the app makes $2 million per day. That translates to 1 per cent of players paying $0.99. The average revenue per daily user comes out to a paltry $0.01. But, in aggregate, all those cents clearly add up to a tidy annual total.

That means the annual revenue per user amounts to a whopping $3.65. Clearly, the individual numbers are not massive. I would estimate that the lifetime value is not going to reach more than $7 – people get bored of games, and new ones keep coming along.

With a relatively meagre lifetime value, it suggests that Candy Crush maintains some pretty lean customer acquisition costs. So, unless the game remains inherently social, popular or consistently talked about, it's going to be quite challenging to keep it profitable.

Zynga famously took advantage of social-user acquisition by running amok on Facebook, exploding to tens of millions of daily users for games such as FarmVille and CityVille. With tiny CACs, the games boasted massive revenues – with similarly massive profit margins. More recently, however, Zynga has been forced to fork out hideous amounts of money to acquire users, which will have an adverse effect on its bottom line.

Going Viral

This is a core component of your economic engine. One of the ways of reducing user-acquisition cost is to focus on referrals. You've probably heard the term 'going viral' a million times before, but what does it actually mean? And how can you make it work for your app?

Going viral is the tendency of an app to be circulated rapidly and widely from one user to another. The crux of the concept is that you make it as easy – and attractive – as possible for your existing users to invite other users to download and use your app. If you can bake in a process whereby a single app download stimulates multiple more downloads, those additional downloads each trigger further downloads – resulting in more downloads much faster, and for free.

Creating these viral loops is based heavily on simple, effective mechanics – and you want to make sure that you understand and take full advantage of all of them. There are five core ones I like to talk about.

1. INHERENT VIRALITY. This means that an app or product generates a very strong network effect. As an individual you won't get much value from it, but, as you share it with friends, you very quickly start seeing bigger and bigger benefits, so you have an incentive to share it. Inherent virality is super-strong with communications-type products such as Snapchat and WhatsApp. Getting the very first users on board is tricky, because without users the network does not yet have value, so it needs to be built up cleverly. Clever sharing mechanics such as simply inviting friends or your entire address book go a huge way to breaking the chicken-and-egg cycle here. Word of mouth similarly drives this type of app.

2. INCENTIVISED VIRALITY. It's a classic marketing idea to offer users incentives to have their friends join. For example, Hailo and Uber both offer credits for you to invite your friends to use the app and take a ride. It's important to make sure that the cost of an incentive is associated with revenue generation (as opposed to just a download), otherwise you won't be driving the most important metric. Another way to approach this is if you have an app or service with 'virtual' value you can offer – the billion-dollar Candy Crush Saga app allows you to earn special bonuses if you invite your friends via Facebook. It was this kind of well-thought-out mechanic that helped propel the game to worldwide number-one status in under 12 months.

3. COMMUNICATION VIRALITY. 'P.S. I love you. Get your free email at HoTMaiL.' That was the inaugural email footer that launched 1,000 similar viral campaigns back in 1996. It was quickly changed to 'Get your free email at HoTMaiL' – but you get the point. It has been used with great success to communicate the usage of a service (or a device) by a single person to all their contacts. Even Apple leveraged this method with their 'Sent from my iPhone' auto-signature. An oldie, but a goodie.

4. SOCIAL-NETWORK VIRALITY. This is all about making it easier for your users to share content – or other relevant information – with their social networks, such as Facebook, Twitter, Pinterest and Google+. There are plenty of ways to do this automatically. For example, when you ask users to create an account with Facebook on your app, you can ask for permission to post on their profile, or on their Facebook feed. Zynga, with its social games, was a powerful – and rather intrusive – adopter of this technique with great effect (though lately this has definitely eroded from its former heights). Zynga was one of the most prolific posters to your Facebook feed – telling each one of your hundreds of friends that you were playing FarmVille, and it even offered you incentives to play yourself. Today, apps such as Flipboard – which allow very simple broadcasting and sharing of your favourite content across all channels simultaneously – is a much better use of this mechanic.

5. WORD-OF-MOUTH VIRALITY. Having your users become vocal advocates of your app is a brilliant channel to propel adoption of your app. Hailo focused heavily on creating a great app for both drivers and passengers to leverage this phenomenon, so, if you've ridden in a taxi in any city where Hailo is operational, there's a huge chance the driver will be raving about it. Square (the register and payment app) is also able to exploit word of mouth because it benefits both merchants and customers, which encourages both sides of that marketplace to start talking about the awesome product and sheer delightful experience that it offers. And yet another great example – this time in the form of a quirky, fun and unique game – is Angry Birds. People just couldn't stop talking about it because it was so addictive (and the great game play, character and sound design definitely helped to a massive extent).

How do I measure my virality?

Virality can be a bit technical, but it is definitely worth understanding it and then measuring it. With the ability to measure it, you have the power to improve it. This can translate into massively lower user-acquisition costs, as well as accelerated user growth.

Your virality is generally measured as a number called the viral coefficient (oddly, it's usually represented by the letter K).

Your viral coefficient, $K = X \times Y \times Z$, where:[4]

- X is the percentage of users who refer or invite other users to download your app;
- Y is the average number of people they invite (over a specific period of time); and
- Z is the percentage of people invited who download your app.

If your K is greater than 1, that means that, for every user you get to download your app, you're getting another K users to download it for free. So, theoretically, if you drive your K to be greater than 1 (which is hard, but possible), then your user acquisition powers itself. In reality, what is does is greatly decrease your average user-acquisition cost – and helps you grow faster. Think of it as the closest thing to a free lunch.

It's worth calculating the viral coefficient of your app: you may discover that you're not exploiting this tactic at all or find that you're actually very close to K = 1 and that a few product improvements will push you over the edge and ignite explosive user growth.

Snapchat is one app that cracked the formula. With no advertising, just very heavy user engagement, it was able to grow from zero users to tens of millions of users in just a couple of years. If you look at its level of growth in terms of photos shared in June 2013, you'll see that Snapchat users were sharing 200 million snaps per day,[5] that's up from 150 million per day in April 2013,[6] and up from 50 million snaps per day in December 2012.[7] If you assume approximately the same referral rate per user, that means you will be able to quadruple the user base in a mere six months. That's viral!

There is another metric that affects your virality: cycle time. This is the average amount of time between when an existing user initiates a 'referral' (invites a new user to the app), and the moment the new user downloads the app. The shorter this period is, the faster you grow. It's worth keeping an eye on, since it may mean spending lots more on servers if you don't.

If you're keen on doing some calculations about how quickly you can generate downloads and new users, then check out this equation from David Skok[8] of Matrix Partners:

$$Custs(t) = Custs(O) \times \frac{K^{(t/ct+1)} - 1}{K - 1}$$

Where t = time
  ct = cycle time

Let's see how powerful cycle time is, and why you might want to optimise your app to take advantage of it (e.g. by sending email prompts to invite friends more quickly).

Let's imagine that your app has a viral coefficient (K) of 1.1 and a cycle time (CT) of 14 days and started with 1,000 users. Within 670 days you could have 1 million users – so not too bad. If you reduced your cycle time to 7 days, then it would take you 50 per cent less time to hit that goal – just under a year at 335 days. It's clear that it can have a huge effect.

Big Data

Right now, because you're still running pretty lean, data is probably the responsibility of an existing marketing manager or analyst and a developer. Once you raise Series B funding, one of the top specialist hires you want is a top data engineer who will be feeding the marketing team mountains of actionable data and metrics. But for the moment you'll have to make do.

As you drive towards becoming a robust, revenue-focused company, you need to ensure that you're making more decisions based on data; you need to take advantage of all the information that you have at hand. While your off-the-shelf analytics solutions are pretty powerful, you will invariably hit a number of limitations. There's a good chance these limitations are going to become very annoying and will start to affect business decisions.

Business intelligence tools such as QlikView give you the ability to generate insight from all the data you have collected on the back-end and pull it quickly into dashboards and reports. Think of it as one level more powerful than your daily analytics tools, and a lot more user-friendly than manipulating raw data with SQL queries (SQL is a special programing language for managing and interrogating data that is stored in relational databases) – and yes, that's meant to sound hard.

At Hailo QlikView allowed us to dive deeper into customer and driver behaviour: the patterns of how long people would wait before cancelling a taxi; which parts of town were more likely to order a taxi, and when; and which people were likely to spend more on taxis.

By having this tool, the marketing team was empowered to ask more intelligent questions and find the answers in the data without using huge amounts of developer time (which is precious, and should be spent on improving the product and performance).

As your company grows, collecting data – and then gathering actionable insight from it – is going to become increasingly important, so it is critical that data should have a key owner in your team – and I would suggest in most cases that falls under the VP of marketing, who is responsible for not only attracting new users, but retaining existing users.

In the next chapter, we investigate the impact that mastering user retention can have.

Chapter 24

Keeping Users Coming Back

Too many companies focus purely on acquiring new users – and unintentionally neglect the power that strong user retention can have on their business. In order to become a great billion-dollar app you are going to have to master both sides of the equation – not only the acquisition (and activation) components but the retention part as well (all of our AARRR – standing, you'll remember, for acquisition, activation, retention, referral, revenue – metrics are critical).

It's a simple concept: you need to be able to master a number of simultaneous and different customer relationships.

So where should we begin? Naturally, let's figure out what our retention rate is, for only then can we tell whether it's good enough, and then, finally, we can do something about it.

Measuring Retention

Remember product–market fit? Well now you don't need to *ask* if customers love your app: you can measure it through retention.

You're going to be really happy now because I'm going to talk about the analytics solution you put in place at the very beginning of your app development process. If you implemented a solution such as Mixpanel, then you'll be able to very easily segment your users by retention rate.

Retention rate is what percentage of your users return and engage with your app within a given period. A monthly retention rate of 50 per cent means after 30 days 50 per cent of the original group of users who downloaded your app came back a second time. That's useful for a number of reasons.

First, say you added in the ability for your users to pay with one click in February. Naturally, you want to see if that improved your retention rate. So, you can check the retention rate before you introduced the feature in January, and compare that with February. If all other factors were equal, and the retention rate was higher in February – hey presto! That new one-click payment system was a winner.

Similarly, you can use retention rate to retarget lapsed users. By digging into your analytics you can target the users who have not opened your app in the last 30, 60 or 90 days. You put together a variety of campaigns you'd like to test – a combination of various messages, offers and calls to action – and then send emails or push notifications to those users. You can then measure the results, and see which campaign brought them back to your app.

It's worth mentioning the concept of cohorts alongside retention rate. We discussed the concept of cohorts in Chapter 19. Cohorts are simply groups of customers who started using your app within a defined period. For example, all of the people who started using your app in January form a new customer cohort for that month. By displaying how often these cohorts return to your app, you can measure whether changes you make to your app make it more or less valuable to customers – and whether they are coming back to your app more as a result.

A good analytics tool will allow you to measure all kinds of key metrics by cohort: acquisition, activation, retention, referral, and, naturally, revenue (the AARRR metrics). By using daily, weekly and monthly cohorts you can very quickly see the effect that product, design or even performance improvements are having on your app.

My Retention is Bigger Than Yours

Fred Wilson, the well-respected venture-capital investor from USV, suggested a retention ratio of 30:10:10 for mobile: 30 per cent of customers should use the app each month, 10 per cent should use the app daily and 10 per cent of the daily users will represent the maximum number of users using the app at any given time.[1] While that's a pretty good rule of thumb, we can do better.

The truth is that retention metrics vary by an app's vertical and business models – so let's have a look at the data. The diagram[2] below shows the retention rates 30 days after downloading an app.

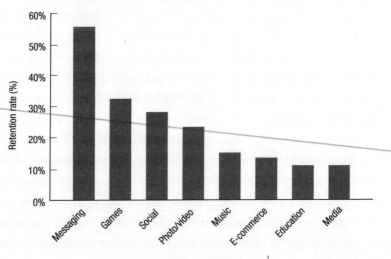

Retention Rates by App's Vertical and Business Model

Messaging leads the pack with a whopping 55 per cent of users, followed by gaming at 33 per cent, social at around 28 per cent and then photo/video at 22 per cent. E-commerce ends up around the 13 per cent mark. So this is a good barometer to see how you're doing against the average.

Hitting average isn't going to get you far. You need to do a lot better than that. To grow and beat your competition, your app needs to consistently retain more than the average percentage of its customers.

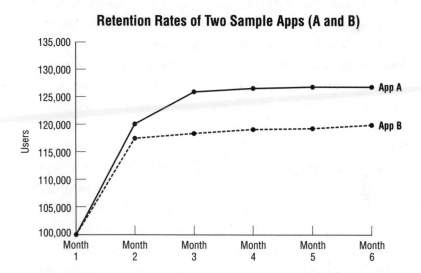

Let's make things a bit more concrete and use an example. Imagine the two apps shown in the graph[3] above are gaming apps. The first, a gaming app we'll call A, has a 25 per cent retention rate, and the lower one, B, has 20 per cent. If you assume they start from the same customer base and keep their retention rates steady for the next six months, A will have 15 per cent more customers by the end of that period. And 15 per cent more revenue if the both are monetised equally. That's a big deal. That's the snowball effect that focusing on retention is going to help you drive.

Taking Action

Now that you can measure your retention rate – and you have a decent idea of what it needs to be to make you competitive – what can you do about it? Basically, you want a system that will allow you to communicate with your users – in a meaningful, and relevant way – that will translate back into increased app usage. That's a win for users, and a win for you.

Historically you'd put in place a CRM – or customer relationship management – system to help you do this. More simply put, a CRM system is a set of rules you define about your users' behaviour – which in certain circumstances triggers email messages, text messages or push notifications.

These used to be pretty complex systems to set up, integrate and operate. Today, they're a lot simpler and friendlier – and more powerful.

Newer and more nimble CRM systems are eclipsing a lot of the bigger systems. One of the best of the new school is Mixpanel. That's right, even though it bills itself as the 'most advanced analytics for mobile and web in the world',[4] it's probably one of the most powerful (though simple) CRM solutions out there.

We've seen many of the features of Mixpanel earlier in the book, but I want to call out two specifically that render it a powerful CRM competitor: people and notifications.

PEOPLE. Mixpanel creates a detailed profile of every user who signs up on your app or your website. No matter what device they log in on, all their activity is recorded against the same user profile. Mixpanel's targeting and segmentation engine makes it easy to see groups of users who have similar profiles, which enables you to understand exactly who is using your product.

Mixpanel even creates an activity feed that puts customer

behaviour into an event-based timeline, so you can follow the way people experience your product. This gives you a powerful way to see where your users get stuck along the way. You can test ways to improve your app by looking at individual customer behaviour. For example, compare the behaviour of customers who recently cancelled to see what similarities they have.

Revenue is a critical business metric. Mixpanel shows you the lifetime value (LTV) of a customer, so you know how much you can spend to acquire more of them while remaining profitable. Similarly, you can drill down and better understand the value of each marketing channel.

NOTIFICATIONS. You can automatically send notifications to your users on mobile and web. You can send emails, text messages and push notifications. The system allows you to target very flexible groups of users, based on any attribute or behaviour you measure. You can then schedule messaging based on any rule you can think of, and then optimise every notification to make sure your users stay engaged. It's immensely powerful.

I don't want to sound like a fan boy of the platform, but it does work wonders. My only criticism is that it does get expensive (but feel free to try to negotiate with them – they are nice people). I have heard a lot of good reviews about KISSmetrics as well. It too, embodies this notion of an analytics-cum-dynamic-notifications platform, so it's well worth checking it out.

The world of CRM has changed a lot from the dinosaur days of enterprise-grade players such as Salesforce and ExactTarget. While those grandfathers have deep roots and tight contracts with big corporations, there are far more nimble, robust and progressive options like the waves of new analytics/CRM players we've looked at earlier.

Hopefully, you've now got to thinking about how to get the most out of your existing users – and how to invest to make them happy. There's also a lot more information about getting started with CRM and the best solutions on the Billion-Dollar App website, mybilliondollarapp.com.

Relevance

The most important quality when you message your users is empathy – put yourself in their shoes. Is the message you're about to send interesting and relevant? Would you like to receive it yourself? If you're sending it only to boost numbers at the cost of a good user experience, challenge yourself to do better. It will pay dividends in the long run.

Some apps do nail their retention messaging very well. Uber messages me once or twice per month with interesting and relevant PR stories. It also frequently emails me with promotion codes that will give me $10, $20 or more off my next Uber journey. It recently improved its referral programme, now offering $20 every time I get a friend to sign up. That got my attention!

As Flipboard is a social magazine, it automatically generates fascinating – and fresh – content. It tends to email me every week with a bespoke round-up of what my friends are reading or recommending on the service. It's interesting, I read it, and I click through a lot.

Tip: make it super-easy for a user either to unsubscribe from your emails or to change the frequency or type of email alerts they receive. If users are unsubscribing, then listen to their actions. Maybe you're sending too many, or they're just not relevant. Rethink your strategy, improve it, test and measure the response. Only by focusing on creating truly relevant messaging will you retain your users in the long term.

Another tip: When you think about emails, remember that 80–90 per cent of the impact is what you write in the subject line, not the body. Make sure there is a simple, clear call to action in the message you're sending. Don't send long rambling emails with a multitude of links and images, but focus on a key action: 'Launch the app now', 'Redeem your free credit', 'Rate the app'. Naturally, this depends on your business (Groupon and Flipboard are designed to present longer, more interesting content and drive multiple, clear calls to action – such as view deal, or share deal).

Alerts

Alert your users about things they care about. People don't mind – in fact they can even get excited – when you alert them to something relevant and interesting. Snapchat alerts me when a friend joins the app; Facebook tells me when I get a friend request or a good friend posts something interesting.

Remember that using push notification alerts is more direct – but also more risky. If you go a bit over the top, users will turn off push notifications for your app. And there's unfortunately no way you can tell if they've done that.

People love alerts about a new message, or new friend requests, but a weekly update about new social news content is something you need to be a bit more careful with.

Turn that Frown Upside Down

There are clever ways to use CRM to turn a negative situation into a positive one. Retention is not just about manually prompting users to come back and say hello. Hailo's world of matching passengers and available drivers is fraught with difficulty: often there are not enough drivers when passengers need them, and sometimes things don't go to plan. Here's how to make the best of it.

Sometimes Hailo's ability to provide drivers precisely when users request them is imperfect. It's at that point the app throws up a 'No taxis available' message. After digging into the analytics, we found that if a user saw this message during their *first* attempt to book a taxi, they never opened the app again. Bummer. It's a logical user response – they probably thought Hailo just didn't work.

As we increased the number of cities we were launching in, we saw this happen a lot more. We wanted to figure out a way to solve this. So we created a rule in our CRM system that would fire an email to a first-time user seeing the 'No taxis available' message. In the email we included a thoughtful message appealing to the user's reasonableness – and, most importantly, offered them $10 off their next ride as an apology.

All of a sudden, we saw a response rate of around 50 per cent to that message. That CRM message is now in place in all our cities, and works like a charm.

Finding Fanatics

There's another benefit of doing all this digging into analytics. It gives you the chance to identify fanatics and cheerleaders. Find the users with the highest retention, the ones with the highest referral rate, or those with the longest usage and the highest revenue. Find them via the app-store reviews, find them on social media. Reach out to them, work with them, forge strong relationships with them. When you create an app that gets people excited, and changes their lives, that's pretty damn rewarding.

Chapter 25

International Growth

As you go through the process of new user-acquisition channels and revenue modelling, the question of international users and markets invariably pops up.

The ability to scale your app around the world is a fascinating and critical challenge, but the best approach to take varies depending on the nature of your business. Some types of apps and business models clearly lend themselves to simpler, faster and more efficient international expansion than others. It is clearly a lot easier, for instance, to make (certain types of) gaming apps go international, than it is for a payment service such as Square or a transportation service such as Uber (which involves all kinds of operational and regulatory challenges that can vary wildly by city and country and require on-the-ground recruitment of drivers).

Now is a good time to ensure that you have put the groundwork in place to be able to pursue international ambitious. We'll go through the basics now, and show how some apps can begin to master international markets early and set themselves up to scale impressively.

Basics in Place for Global Domination

Let's have a look what needs to be in place before you decide you want to be a Master of the Universe and go after all these new countries.

The very first thing you need to do is make sure your app can be easily translated into foreign languages. The mechanics of doing this breaks down into a few sub-areas. Let's look at each in turn.

STRINGS FILES. First, make sure that the text content – i.e. the words in your app – is actually contained in a 'strings file'. This means that all the words used in your content, on buttons, on forms and in the navigation are stored in a single file that can be translated and used within your app. Doing this means that, as soon as a user changes their device settings to display a language that you have included in the strings file, that file will be loaded to the user's app and, hey presto, your app is in Japanese! (It was a wild day when I first saw the Hailo app in Japanese.)

Be careful, though. While the concept of strings files is standard practice, making sure that the Android and iOS apps use the same strings file – and therefore avoiding the need of translating both apps separately (and doubling the complexity in your day-to-day life) – is often something that happens too late. It took us a month to sort this out!

At Hailo we messed this up. Having started with an iOS-only strategy for our driver app (which was also a lot more complicated, and therefore had tons more text in it), we didn't enforce enough discipline when we started on the Android version of the driver app. We wasted a good couple of months having the iOS and Android engineers sync up two very different naming conventions to make translation actually possible.

LANGUAGE TOOLS. Once you actually go down the route of supporting more than one language, you're actually forced to be rather methodical with all your changes: the moment you add a new feature in English (or just tweak some content) you need to make sure you do it in all your other support languages. Luckily, someone thought of a solution for this issue and there are now two pretty good tools out there. WebTranslateIt (WTI) is a great one. Another is called Smartling – and it's gorgeous. But, unfortunately, it's eye-bogglingly expensive and just doesn't add much functionality over WTI. These tools allow you to quickly and centrally manage all your translations – for your website and apps – and allow third-party translators to quickly translate any changes your product and development teams make. It is integrated into the software-development process.

TRANSLATIONS. When you have the entire translation infrastructure in place, you then actually need to have the content in the string files translated. There are a number of firms out there who specialise in this – and it's not something that's as simple as it first appears. You need translators who are familiar with apps and with your industry and service, and clever about their translations. Remember that you're working with limited space on an app, so every *character* counts (and your translators need to appreciate that). The great ones we worked with at Hailo included Beluga Linguistics and Lingo24.

Angry Birds Take Flight

Angry Birds launched in the App Store in December 2009. It may seem absurd now, but it was a complete flop. The team behind Angry Birds were called Rovio. The team were based in Finland. From the beginning they knew they needed to get into other, bigger, markets if the game was going to be a success. From the get-go they focused on growing internationally. They couldn't get any traction at all in the UK and US markets – not that surprising, given that the App Store competition was brutal. The team paused to rethink their strategy (needless to say, they were pretty thick-skinned: they released 51 games *before* Angry Birds). They turned to cunning, rather than persevering with a full-frontal assault.

'We tried to get a following in the smaller nations,' says Matt Wilson, head of marketing at Rovio. In these territories, Angry Birds could gain visibility with far fewer sales. For example, just a few hundred purchases took the game to number one in the Finnish App Store.

Rovio duplicated that strategy in other countries and hit the number-one spot in the App Store in Sweden and Denmark, then Greece and the Czech Republic. By employing this strategy it collected about 40,000 downloads in smaller markets – positioning it well to enter the more lucrative UK market.

Next, Rovio inked a partnership with Chillingo, an independent publisher that helps app games developers not only distribute their games but also tweak the games to help them get traction in different markets. It has a great relationship with Apple, and has helped other top games such as Cut the Rope, Feed Me Oil and Contre Jour break into the big time. It was credibility in these smaller countries, and this partnership, that put Rovio in a position to pitch to Apple about getting on the front page of the App Store. On 11 February 2010, Apple featured Angry Birds as 'Game of the Week' on the front page of the UK App Store.

This was a huge opportunity for Angry Birds (when Hailo was similarly featured, it received an influx of downloads in the high tens of thousands on the first day alone), and Rovio recognised that it had to make the most of this exposure to cement its top App Store position. It did this by simultaneously releasing more than 40 new levels for the game, and then releasing a freemium version (i.e. free to download, pay to get more levels). Rovio was agile, and delivered everything within a few days.

The strategy worked. The single feature propelled Angry Birds from around 600th in the App Store to number one. Only two months later, in April 2010, it hit the number one spot in the US App Store as well. Rovio was able to deliver this kind of agility because it was cunning about its app product design. It didn't have any tutorial, very little language to translate, and a highly intuitive interface that completely cut across language and culture.

You need to be daring in terms of your approach, your design. Fortune favours the bold – but also those who do a huge amount of user testing and iteration. Then they know their product works.

Building an Uber Empire

Some apps were designed from the ground up to be easy to adopt, use and enjoy – and be language- and culture-neutral (there are massive benefits to this, clearly). On the other end of the spectrum are apps such as Square, Hailo and Uber.

These three apps operate in the highly regulated areas of financial services and payments and transportation. It is immediately obvious that there is going to be a big amount of legwork not only to start operations in one of these areas, but, more interestingly, to expand at an exciting rate. Let's see how these companies have each approached international expansion in a different way.

Uber started out as an app to hail and pay for black cars (the US version of minicabs or the big London minicab firm Addison Lee) in San Francisco, where taxis are in very short supply. The idea worked, and soon spread to more than 60 cities around the world.

Travis Kalanick, Uber's CEO and founder, adopted a bold strategy that has clearly worked – but certainly isn't for the faint of heart. According to him, Uber is a 'cross between lifestyle and logistics'.[1] But Kalanick wants Uber to be an 'instant gratification' service that gives people what they need, when they need it, whether that's a ride in a black car, taxi or some other delivery (Uber has delivered all kinds of things, including ice cream, roses and even helicopters and boats).

But the real vision might be what one of Uber's investors, Shervin Pishevar, recounted during a recent interview:

> 'Uber is building a *digital mesh* – a grid that goes over the cities. Once you have that grid running, in everyone's pockets, there is a lot of potential for what you can build as a platform. Uber is in the empire-building phase.'[2]

So how does Uber go about building this empire?

Uber targets the cities it wants to launch in and then goes about finding GMs – general managers – who are tasked with setting up and running the entire operation in that location. In recruiting these GMs, Uber tends to target ex-investment-banker-with-an-MBA types (I have met a number of them) who characterise the company's personality. They are essentially entrepreneurs – with or without local market expertise – who go in there and make things happen. The goal is to get the service up and running with as few as 25 drivers. In fact, Uber revealed that, during the recruitment process for its GMs, it asks them to present both a launch plan and then a six-month operational plan for the city they wish to manage. That's a pretty efficient – and to-the-point – recruiting process.

On its website, Uber describes its GMs as people who have 'a fearless, confident and collaborative personality' and possess 'a pragmatism that building a new, city wide transportation system requires'. The GM is entirely 'responsible for managing people, regulatory oversight and ultimately the development and growth of [the] business'.

The GMs receive all the collective experience and wisdom of the rapidly expanding company through their 'Playbook'. The Playbook comes directly from the CEO and executive team and presents an extensive collection of strategies for overcoming obstacles on a local level, and includes everything from how to launch and operate social-media channels, to how to throw a launch party, to how to recruit key employees (such as D-Ops, or driver operations staff, and CMs, community managers) and deal with regulators. This last point is particularly important.

As it expands, Uber has run up against regulatory and legal issues. Governments – around the world and in the US – are very concerned at the pace of change and their ability to control it. Numerous startups are succeeding at causing governments similar problems. Airbnb and its new-fangled approach to monetising your spare room or apartment drew ire from the New York Attorney General, who said the people renting their properties are breaking the law.[3]

With a fresh injection of cash – a quarter of a billion dollars in August 2013[4] – Uber is now very well positioned to escalate its operations, move even faster and fight legal battles. In many locations, it seems that Uber is slowly changing perceptions and is even seeing the laws change in its favour – something previously unthinkable.

It's also worth noting that Uber keeps its technology pretty damn simple. The platform and app to deliver job requests to its drivers, for instance, is very simple (I have had a good chance to play with it). This simplicity is very important because it allows quick translation and then dissemination across its various markets. Additionally, Uber distributes its driver app preinstalled on an iPhone (owned and paid for by Uber) that is given to its drivers free of charge. And, while this is a huge capital expense, it is the easiest way for a driver to get up and running with Uber – without complexities associated with downloading, installing, and then creating an account on the service.

The real key to this savvy international expansion strategy is to target the right geographies, put in place a GM who knows what they are doing, support it with a skeleton team to keep the costs down, and then ink deals with the local car transportation companies. With just 25–50 cars on the ground, Uber can launch its service in any given city. While this won't provide the best service or coverage, the company realises that, by appealing to the tech crowd first, it will get a bit more forgiveness as it bolsters and tunes the types of cars and drivers on its network. It's a model that's been shown to work.

Hailo Goes Global

Hailo – the taxi app – has many things in common with Uber, but also many differences in terms of its approach to international expansion. It was much more focused on developing a deep footprint in each one of its world cities than on chasing a huge number.

Its very first foray into global expansion was jumping across the water from London to Dublin. Hailo recruited brilliant local GMs. However, its focus was not on building a network of ex-investment bankers and B-schoolers to grow the business, but rather seasoned operators with talent for people management and operations.

Its initial Dublin GM (and now GM of Europe) was a good friend of one of the cofounders (and actually a CEO at a competing business) who was tasked with putting everything in place in the new market, including recruiting the best taxi drivers to lead the local driver team (Hailo doesn't recruit salespeople, but rather prefers to work directly with taxi drivers to tune the driver app and then get the word out there and sign up drivers to use it) and hiring social-media and marketing staff.

The big next step for Hailo was establishing a US beachhead, and for a number of reasons New York – possibly the most iconic cab location in the world – was chosen. Hailo's CEO Jay Bregman is an NYC native, so the process of getting set up there was quick and very much enabled by Hailo's investors as well.

Unlike Uber, Hailo thrives in the highly regulated taxi market. Part of the Hailo strategy is to work very closely with regulators to ensure every part of its service adheres to the applicable laws in each location. What was surprising – and also depressing, because of its complexity – was the wide variance of customisations that had to be made to both its passenger and driver apps to have them work in a given city. In most cases the regulations differ by city – not just by country – leading to a huge number of variants of Hailo's apps.

One of the big projects that the team delivered in 2013 was the ability to customise its apps on the fly by creating a Web interface to allow the team – and especially the GMs – to configure the app based on the needs of their respective cities. All this could be accomplished without doing a new release of the app (the app was smart enough to call back to the server and grab a new configuration every time it was turned on), meaning that all kinds of features – everything from the range of languages, currencies and payment options supported by the app (in some cities Hailo is obliged to allow payment by cash) to the regulatory requirements around city-specific taxes, surcharges and even driver-specific terminology and payment frequency and options – could be configured via simple Web interface.

While this was a substantial investment in product, design and engineering time, Hailo now has an unparalleled platform that will allow it to scale to hundreds of cities simultaneously – and still be able to adhere to the relevant laws and requirements in an agile way.

Scaling Mount Fujii

Japan was always a target country for Hailo. It's one of the biggest taxi markets in the world; the Japanese spend upwards of $30 billion each year to be driven around in immaculately maintained cabs. It's also the only place in the world where the calibre and knowledge of drivers rivals those of London cabbies.

Finding the right person to help launch Hailo in Japan was critical. We knew we needed a local, we knew they needed the political experience and acumen to navigate a rather tricky landscape, and the company needed a leader who would not only get things up and running quickly in Japan, but also lay the groundwork for rapid expansion in the rest of Asia.

At this level, headhunting can be costly and complicated, and, funnily enough, it was Accel Partners who came to the rescue once again. It was through their vast network that they introduced us to Kiyotaka Fujii. Native Japanese, but with strong experience in the West, Fujii helped SAP and LVMH launch in Japan with great success.

It was Fujii who enabled us to launch Hailo in both Tokyo and Osaka.

Square in Japan

Square trailblazed the payment app space incredibly well – so much so that there spawned a tidal wave of clones across Europe. Sweden launched iZettle, Germany's Rocket Internet rolled out Payleven, SumUp popped up in the UK, and PayPal pushed its mobile merchant payment app on both sides of the Atlantic.

So, in typically clever fashion, Square shunned Europe and instead headed for Japan. You may think that this is a bit peculiar, but on further inspection it actually stacks up as a pretty wise first move outside the US and Canada.

Europe is a pretty tough market for mobile payment apps. It's not just because the market is fragmented, complex and fraught with anti-competitive national governments, but it's also more demanding on credit-card transactions – with mandatory PIN codes for debit and credit cards. As a result, regulation requires that this new wave of payment apps must require the user to input their PIN to execute a payment. Since that can't be done on the merchant's smartphone (for security reasons) it has necessitated an entirely new piece of hardware: a standalone card reader. And so the problems begin. This not only undermines the elegance and simplicity of Square's plug-and-play mini-reader, but it introduces a whole new level of complexity, as the European card readers require a Bluetooth connection and a not-that-simple setup process.

It's a first-class product failure.

Upon launching in Japan, Square was competing only with PayPal, which wasn't seeing much traction. And Square was able to launch with its iconic, sleek mini-card reader, affording its Japanese customers the same exceptional user experience.

'I am honoured to introduce Square to a country with a rich history of design, innovation and tradition,' said Square's CEO Jack Dorsey. 'Square shares the same values and attention to detail in our products. Our tools are made to enable business owners to create a delightful, seamless experience for their customers. I look forward to Square assisting in Japan's continued economic growth and entrepreneurship opportunities.'[5]

With no local competitors, Square is partnering with Mitsui Sumitomo Bank – a huge national player. While it is charging merchants 3.25 per cent for transactions[6] (more than in the US and Canada), the rate is much lower than the 5 per cent PayPal is charging.[7]

Square's association with the iPhone will also help it in Japan, where Apple's smartphone is still beating out Android devices. The iPhone makes up 66 per cent of sales there, compared with Android's 32 per cent share.[8] It's worth noting that Square supports a wide variety of Apple and Android smartphones and tablets, which means potential for faster expansion.

Go East, Young Man

China is an incredibly fascinating market – I had a chance to dip my toes in while at Hailo. A great friend of mine, Hugo Barra, is quickly becoming an expert. He left his post as head of product at Google's Android in 2013 to move there and work for Xiaomi,[9] a young but already billion-dollar mobile handset manufacturer led by CEO Lei Jun, affectionately known as the 'Steve Jobs of China'.

Hugo paints a fantastically interesting picture of China.[10]

- There are more than 618 million Internet users in the country – and that number has grown by an amazing 50 per cent from 2010 to 2013.[11]
- Chinese Internet companies boast massive user numbers (QQ – the chat service – has 500 million; QZone – the social-networking arm of QQ – has 600 million monthly active users; WeChat – a messaging app like WhatsApp – has about 270 million; MoMo – a social dating app where you talk to strangers who are near you – has 100 million users).
- E-commerce is booming. Taobao is the country's leading shopping site – it is twice the size of eBay and Amazon combined. Not bad.

The Chinese also have more money to burn – their disposable income has tripled over the eight-year period from 2005 to 2013.

Due in part to this massive technological adoption, that country now has at least 122 billionaires (only the US has more).

There is clearly a massive amount of potential in the East. While it may not be the easiest place to do business, it's worth thinking about. There are massive opportunities – and they are only getting bigger. With more than 8 million university graduates every year (that's more than the US), their skilled workforce is growing like crazy, and may well set its sights on the West. Tencent – the company that owns QZone and QQ – is rumoured to have invested in Snapchat.[12]

So it's worth remembering: the sun rises in the East.

Chapter 26

Growth is a Bitch

So far this section has been all about growing your business, and, while this is hugely exciting, it can also present challenges for your company. You need to think about scaling your organisation, and that requires preparation. You're going to need structures, systems and processes in place. You may think it's boring – but it will need to happen.

Struggling with growth is one of those problems you want to have. During this growth stage the team at Hailo grew from 25 to 100 people (it took us a mere nine months from closing our Series A to closing our Series B). The growth was amazing – but, in order to make it deliver value, we needed a very clear organisational growth strategy.

Hire for Tomorrow, Not Today

At this critical junction, where you can see the day when your app is going to grow into a big company, you need to start exercising some forward-looking vision. At Hailo we had a pretty good vision about where we needed to hire people, and put in place a plan accordingly. We had a pretty ambitious list of major world cities that we were targeting for launch, so the priority was getting in great managers to get the operations going in those geographies. While they were pretty senior hires, these managers needed to be capable of building their own city teams – and recruiting for them began six or more months in advance of an anticipated city launch.

At the same time, on the technology side we knew that it was actually the complex centralised platform that required more engineering effort, rather than specific development on the app side. So we started ramping up on engineers who would work specifically on those systems.

Two things help along the way – and should constantly be in the mind of the CEO and COO. First, you need to have a clear idea about what your org chart looks like. Having a clear vision and structure for which people you need running your company is a prerequisite to getting the right people in place. Don't shy away from this. Put an org chart together – and then revisit it quarterly to adjust it.

Once you have a version of your org chart, you can highlight the key hires that you need to make. It becomes increasingly important to know precisely where experienced hires fit into the organisation, and to have someone who is responsible for their onboarding and management. In the early days of a startup this matters less, but it is important to formalise.

One of the most important things a CEO can do at this point is to be continually recruiting. Whether it's seducing a spectacularly talented software developer, a VP of marketing or a general manager, a key role of a CEO is to be always networking and scouting for talent. As a company gets seriously ready to scale, grooming potential big hitters to come and join the team should take at least 20 per cent or more of the CEO's time.

Goal Setting

As your organisation takes shape and as your hiring plan starts delivering results – and you grow from around 30 people and start hitting 50, 75 or even 100 – there are some organisational things you need to put in place. Many of these are easier when you're smaller, so think about them early.

One of the most important is about setting – and hitting – business goals. Most companies are not particularly good at this, unless they have seasoned serial entrepreneur cofounders. At Hailo we addressed this late, and could have spared ourselves a lot of meetings and additional work by implementing it earlier in our lifecycle.

One of the best frameworks for how to set, measure and hit business targets was invented by Andy Grove of Intel. Grove wrote a famous manual called *High Output Management*. It can be distilled into two core questions:

- Where do I want to go?
- How will I know I'm getting there?

They are two fundamental questions that every business needs to answer. A framework that came out of his book and Intel was OKR – or 'objectives and key results'. Famed venture capitalist John Doerr popularised the framework, which is now actively used in companies such as Google and Zynga.

The idea is that the whole company and every team has one objective and three measurable key results every quarter, and, if you achieve two of the three, you achieve your overall objective; if you achieve all three, you've killed it. It is a good, simple organising principle that keeps people focused on the three things that matter – not the ten.

As Mark Pincus, cofounder and former CEO of Zynga, has said, 'It's about making everyone the CEO of something' – from the management team through to every single employee. The principle is a good one: everyone in a company should understand what the company's objectives are and be responsible for helping to achieve those objectives, in whatever capacity they're able.

While the OKR planning process starts off as a top-down approach, when all the OKRs have been established for a quarter they should be made accessible to everyone in the company to review whenever they choose. One good approach is to have a shared Google spreadsheet – or something like a company wiki – with pages for each team and members (management, product, development, marketing). By sharing everyone's OKRs with everyone else it's a powerful statement of the leadership's commitment to transparency.

Friday Updates

When Hailo started holding a Friday afternoon 'all-hands meeting', it elevated the company to a new level of openness and efficiency. People had their heads down all week, and they knew they would get an update on everything – from performance, to hiring, to challenges – every Friday afternoon (and beer as well). As it grew, it invested in high-quality video conferencing gear so that its teams in North America, Canada and Europe could all be part of the same conversation – and feel like one company.

The same process was institutionalised at Google from its earliest days with its TGIF ('Thank God It's Friday') meeting at 4.30 p.m. every week. In addition to getting a frequent update from the head of each team, it was also a platform where employees were empowered to ask tough questions in a public forum knowing both that they would get an answer on the spot and, more importantly, that there was no such thing as an inappropriate question. An environment of trust and transparency is key to creating a workplace where people can focus on delivery rather than worry about how the company is doing or distracting politics.

Square's Jack Dorsey implemented another interesting approach to promote a transparent environment. Every important metric is emailed on a daily basis to every employee of the company; everyone knows precisely how the business is performing. Another rather distinctive policy is that, for any meeting with more than three people that lasts thirty minutes, the meeting notes need to be emailed to everybody in the company. Naturally, that leads to some overflowing inboxes, but it's something that seems to be working at Square.

Growing your business to the $100 million stage is all about scaling, and the first part of that process is making sure you have scaled your internal communications effectively. It's not necessarily more work – but it is a new task. And it is super-critical.

Pay Attention to the Small Things

When you're building up a technology company there are plenty of mutually advantageous things you can do for your employees that don't cost you very much time or money. For example, giving your employees the chance to attend conferences achieves a variety of goals. Not only does it help to develop their professional skills, but it also leads to great networking, which also leads to better channels of recruitment. And, frankly, we all know that techies love geeking out at conferences – and making sure they are aware of what's bleeding-edge.

For some members of the team these events will afford them opportunities to talk on behalf of the company – thus raising everyone's profile in the tech community. Again, this kind of outreach allows your employees to become real leaders in a domain they love, and it gives your company – as well as the app – all kinds of free publicity.

While you're still in this high-growth phase, these opportunities are a great surrogate for actual 'career development' plans or progression plans. At Hailo we sent our iOS engineers to the WWDC (Worldwide Developers Conference) every year to see the latest and greatest from Apple, and, similarly, our Android team to the Google I/O conference. These kinds of trips return highly motivated engineers who want to try out brand-new things in their own app. Priceless!

Growing Your Product and Development Teams

In the previous section we talked about the team responsible for building your app: product managers, designers, developers and QA (quality assurance – or testers). At this stage of your business you're hopefully still pretty lean – but some teams may start reaching a size of 20 or 30 people. The vast majority of your team should be focused on product and development.

In reality, it is with about this size of team that inefficiencies and friction can start to creep in. During this high-growth phase, Hailo encountered a few bumps: whereas it had been easy to deliver new versions of the app every week, the pressures of delivering integrations with new payment systems and reporting systems; creating new email alerts; tweaking the back-end systems that monitored the real-time locations of our taxis; and supporting a new mobile platform (we added Android) all started to become too complex.

The challenge was twofold. First, there was an imbalance in resources: the product team, Rob (the designer) and me didn't have enough bandwidth to provide all the design and wireframes the developers needed. Second, there were conflicting views about what people to hire, and how those people should be best organised to deliver the software we needed.

What I know now (thanks to Marty Cagan, a Silicon Valley product guru who's helped countless product and development tech teams grow into efficient operations) is the following:

- All startups have very similar growing pains as they grow their product and development teams – and the final solutions are very similar and proven to work.
- Most companies end up in a 'product-centric' organisational structure, which means that there is a product manager with a development team and testers dedicated to each one of the core product areas: the mobile apps, the customer-facing website, the back-end systems, the admin and reporting systems and the services layer (which is the API, or application programming interface, we talked about earlier).
- These self-sufficient teams have all the skills they need to design, develop, test and then release software to real users out in the wild. A good rule of thumb is one product manager for eight to ten engineers, one designer for every two product managers, and one tester for every product manager.
- As your team grows, make sure that you hire product managers before you hire engineers. This ensures that you have a backlog of features for engineers to work on – and there's someone there to guide the engineers well as they work.

To make sure everything is functioning smoothly it's very important to have the optimal ratio of people – and roles – in your development teams.

Scrum Masters

One of the most helpful things I learned at Hailo – which I was completely unaware of at my previous startup – is the difference that a truly great scrum master can make. So what is a 'scrum master'?

We talked earlier about the 'agile' app software development. 'Scrum' is the most common way to flexibly and holistically develop products in an agile way where a development team works as a unit to deliver a common goal (like build an app). The scrum master is the person in charge of helping the developers focus on building software, by removing all distractions and dependencies. In one sense they are a very focused project manager who works with the product team directly and as the main interface with developers.

When you get to a decent-sized development team – probably around 20 or more engineers – then you can benefit from a scrum master (I am not talking about an average project manager, but someone who is a trained – and certified – scrum master, and who loves making development teams more efficient). My epiphany came when at Hailo we hired a truly great scrum master, and saw improvements not only in developer happiness immediately but also in how many features we were delivering.

Keep in mind that scrum masters vary wildly in experience, ability and passion – and you get what you pay for. But at the right time, when your development team is growing quickly, you can reap some big efficiency and moral improvements. A good ratio is one scrum master for every 12 or 15 engineers.

Big Finance

As CEO, one of your key responsibilities is keeping an eye on cash flow, but from the outset of your business you need to have a clever accountant on board to make sure your finances are in order and that you're filing all the proper accounts. Typically, a good accountant (ideally with a bit of startup experience) will be adequate until your Series A investment.

After that it might be a good idea to think about bringing the finance function in house. If your business model is complex, or it generates significant international revenues, or if your business depends on multiple partners and has complex commissions, it will increasingly make sense to have someone full-time and on the payroll managing this.

So when should you hire that elusive director of finance? A good indicator is whether you and the executive team can easily access all the financial information you need in a timely fashion. Do you have daily revenue, cost and profit numbers? Do you have a solid forecast of monthly sales and cash-flow numbers? If you're not sure, you need to get someone on top of this.

One option is to outsource your CFO (or chief financial officer). You can think of this as upgrading your accountant to someone who has all the skills of a CFO but will cost a fraction and still provide you with all the financial reporting, modelling and reports you need to effectively run your business. I list a number of good firms in the US and UK on mybilliondollarapp.com.

One of the cofounders at Hailo (Ron) was an experienced finance and private-equity guy, so the financial models for the early investment rounds were in great hands. But, as we started to expand our international operations (Ireland, Canada and the USA came quickly after the UK), the overhead of managing all accounting, cash, banking and other financial complexities justified hiring a VP of finance. We needed to get the Series A cash in the bank to afford the guy, but, once the funding round was complete, he was our next full-time hire.

Despite the cost, having a tight grip on financials is going to help scale your company. As with other leadership roles, Hailo's new VP of finance was able to recruit and manage his own team across a variety of territories. Delegation is key to efficient expansion. This person will also be instrumental in preparing all the numbers and financial models you need to create a compelling picture to investors – and secure future funding should you need it.

Chapter 27

Money for Scale

The ideal scenario at this point is that you've been working hard growing your users and tuning your revenue engine, and it's been working. The best case is that revenues are growing nicely, and you've nailed a profitable model, which is highly likely if you're focused on gaming, where the margins are very high – think Clash of Clans, Angry Birds, Candy Crush Saga. But, if you're focused on e-commerce or marketplace, your margins are more likely to be in the 3–20 per cent range (think Hailo, Uber or Square). And, if you're in an 'audience-building' game (such as Snapchat or Instagram), you won't have any revenues at all.

So in most cases – to keep the momentum and grow bigger, faster – it's now time to start looking for more investment. On average, you're going to be looking to raise anywhere from $10 to $25 million if all is well. Hopefully, if you've hit a strong valuation at, or north of, $100 million, you might get away with giving away only 10 per cent of your company.

Timing

The typical timing between closing a Series A funding round and landing a Series B round is around 600 days.[1] You know from earlier in the book that it takes three solid months to close out an investment round when things are going well (and potentially longer if things are not going so well, but we'll delve into that later).

So you have just over 18 months to demonstrate that you have tuned your business model and started to grow users, and therefore have a credible strategy in place to scale your business. Your marketing will have to be well under control with reliable user-acquisition costs and a robust model to demonstrate the expected lifetime value of your users.

Along with that, you will have a clear idea of what your dream organisation looks like – and who your key hires are going to be. All you need to get in place is some additional cash to accelerate the process.

The better your app is performing, the more options you'll have when it comes to securing Series B investment. When you talk to investors you will have a choice about whether you take money to focus entirely on marketing and building your team or even to acquire smaller competitors to enforce your dominant position. The average Series B valuation for all startups was around $27 million in 2013.[2] What you have to remember, however, is that the best companies – the category leaders – break away from the average at this point. They don't just double or triple their valuation from the Series A round: they increase it 10 times or more.

Let's have a look at what our friendly Billion-Dollar App Club members were doing when they were at this stage.

SNAPCHAT. These guys are the exact opposite of the self-sufficient WhatsApp. At this stage their app had no revenue stream and their massive growth was generating huge technology costs. Luckily, their growth and popularity has meant that they are able to pick from the cream of the crop of investors. A mere 13 months after their Series A investment, they closed a $60–75 million dollar[3] Series B round (at an $800 million valuation) – meaning that they gave away only 10 per cent of their company (that's not bad at all). They used the money to invest in technology infrastructure and make a big COO hire.

UBER. This company raised $37 million in its Series B round at a valuation around $300–350 million. The company also used the money to hire Kees Koolen, who grew booking.com to $9 billion in revenues as the company's number two, as COO.[4] Koolen then helped the app through a phase of massive European and international expansion.

SQUARE. On the other hand, Square needs significant amounts of cash to scale. Its business model yields tiny profit margins, which means it needs *huge* scale to make the business work. It raised a $27 million Series B (and much larger follow-on rounds). It was rewarded with great valuations because of its performance – but, when your gross profit is 2.9 per cent (compared with Rovio's 70 per cent), clearly you need to get cash from a source other than your own revenues to grow!

If your app is currently generating significant profit and you've been clever about keeping your operating costs down, it's also worth considering whether you need to secure any investment at all at this stage, as it will mean giving away a stake in your business. WhatsApp is a great example of high performance and keeping control of your company and equity. It has been a lean and profitable company from the start, meaning that the founders accepted only one round of strategic financing (for $8 million) about two years after its launch from the tier one venture-capital firm Sequoia. Koum, the CEO, said it was for strategic reasons. Presumably, it was to cement the company as being worthy of a seal of approval from one of the most prestigious investment firms in the world.

Similarly, Rovio's Angry Birds haven't needed any more investment after their Series A financing round because they have a great business model – a combination of a freemium app, in-app purchases and brand licensing – and great financing management. In fact, they announced $200 million in revenues for 2012, $71 million of which was profit.[5] Who needs investors with that kind of profit margin?

Which Investors to Approach

Building a technology company is pretty tough, and being successful at each stage requires a different set of skills. On a similar note, professional investors tend to also specialise, depending on which phase of a company's lifecycle they focus on investing in.

The diagram on page 322 gives a feel for the types of investors who specialise at each stage. At the Series B stage you're going to have the opportunity to talk to a lot of the classic and well-known venture-capital firms, such as Accel, USV, Sequoia, KPCB, Index, Atomico, Greylock and Andreessen Horowitz (there's a comprehensive list on mybilliondollarapp.com).

Knowing whom to approach is very much a function of the network you've been building yourself, and how well you've been performing. At this stage, if you're growing well and hitting good numbers, then VCs are also going to be reaching out to you as well.

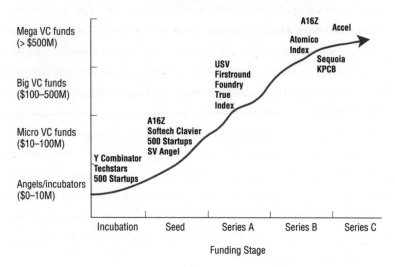

VC Firms by Funding Stage

Going Flat

But what if things are not going quite that well at this stage? What if you aren't able to create a product that is loved outside your core market – or in any new market? What if a competitor wipes out your profit margins? Or, perhaps it's just taking a lot longer to figure out how to grow your user base? You still need cash to stay afloat.

There are a few options open to you, but they're probably going to cost you a lot more equity than would be the case if things were going well. In this case you're in a situation called a 'flat round', which means your valuation has stayed flat since the last time you raised money; or, worse, you're in a 'down round', where your valuation has actually decreased.

In both these scenarios, you're going to have to use all your sales skills to try to convince investors that you will do anything required to revise your strategy to break out of this negative streak and get back to growth. Flat or down rounds occur in about 35 or 40 per cent[6] of venture-capital financings, so talk to your close VC friends and find out the best options.

Another option is a bridge loan.[7] In this case, one or more of your investors gives you additional cash that will serve as a bridge to an event in the future. At the agreed point in the future, you and your investors will see if the valuation has improved based on the results you've achieved. This is the best option you're going to have, as it doesn't create a depressing scenario whereby you have to part with a big chunk of equity at a flat valuation. Similarly, it makes sense for investors to ensure their entrepreneurs are kept motivated – and supported – in what is a particularly trying phase of building a company. I agree with Fred Wilson of Union Square Ventures when he says that '. . . VCs and angels are supportive of their portfolio companies well beyond what a hard-nosed rational investor would be.'[8] So do explore various options should you ever get into a tough position.

The journey is always going to be full of ups and downs. The best entrepreneurs are the tough ones who persevere – but adapt quickly – and try out every possible option. So don't be disheartened. As Apple cofounder Steve Jobs once said, 'I'm convinced that about half of what separates the successful entrepreneurs from the non-successful ones is pure perseverance.'

Summary

This step has been a transition. It's been about getting everything you need in place to scale your business. It marks the point from being just an app – and experimenting with product, users and a revenue model – to being ready to scale into a fast-growing, large and sustainable business.

If you've made it to this stage you've conquered some of toughest challenges out there: focusing on a huge addressable market, reaching product–market fit with your app, putting together an organisation that can demonstrably add users and generate revenue profitability. It's now a question of expanding that brilliant core into a thriving company.

This is an important stage for any app – and any company. What you need to be very aware of, though, is that you judge your performance accurately, and that you really are ready to scale up. Scaling up prematurely is fraught with dangers – and the desire to move to that next stage will be great.

In the next section we'll dive into evaluating whether you're truly ready to scale – and go to the next level.

STEP 4

The Five-Hundred-Million-Dollar App

Scaling Your Business and Raising Series C Funding

Going from a Hundred-Million-Dollar App to a Five-Hundred-Million-Dollar App

App

- At this stage your app continues to receive great reviews, you have a clear path to receive and analyse all user feedback, and you have honed the process of continually trying and testing new features to improve the core user experience and your conversion funnels.

- On top of that you now need to leverage the network effects, scale and brand of your app to drive further big innovations and cement your app's leading position.

Team

- Your team continues to grow – and at a faster rate than before. As you approach the $500 million stage you could hit a couple of hundred people (though this will vary depending how profitable your business model is).

- As you scale, you're most likely going to feel the need to hire experienced people who have done it before in key roles. There's a good chance that an experienced COO (chief operating officer) will put in place the rigour and processes you need to grow efficiently and smoothly.

- With robust revenues (and new funding) you will have the option to bring in people to help scale other departments, and focus on attracting specialist hire to further enrich your engineering, data, marketing and design teams.

Users

- Your user base of millions should be growing steadily (or thousands if you're focused on enterprise customers).

- Your user-retention strategy is not just in place, but is now actively keeping users engaged, and hopefully helping to drive referrals as well.

- User acquisition has now become systematic and predictable – and the cost of acquiring a new user is constantly being driven down in new and inventive ways.

Business model

- As your app improves, you're continually improving the lifetime value of users – keeping them coming back time and time again.

- At the same time you're constantly thinking about how to defend your business model, tune pricing to keep it competitive and maintain differentiation from your competition.

- Your company is becoming increasingly profitable with revenues growing from the tens of millions, and now creeping into the hundreds of millions.

- And you constantly have a eye out for the product innovations that are going to inject fresh revenue (and profit) streams into your business.

Valuation

- Everything is different now. At this stage – as you scale – you're no longer going to be treated as a startup. You're now a real company and you need to make sure your financials are being managed by a real CFO.

- Your valuation is no longer simple to calculate. It's going to be based on how strong your financial performance is (how fast you're growing, and how profitable you are) and how unique (and defendable) your business is.

Investment

- In an ideal world, with a profitable business model in place, you can use the profits you're generating to fund further growth.

- If you need more funding to grow faster than the competition (or even to acquire competitors), or go international in a big way, then the more profitable and fast-growing your company is, the more likely it is that you will attract investment from the top investors.

Chapter 28

Shifting Up a Gear

It was an amusing sight: thirty-odd scruffy-looking developer types pushing tired-looking office chairs on the footpath alongside the River Thames in London. The chairs were piled high with computer monitors, books, printers and various office odds and ends. The group snaked from the HMS *President*, Hailo's former office, to the revered Somerset House, some 400 metres away.

We had finally outgrown our nautical offices – and had just closed our Series B funding. It was a beautiful end-of-summer afternoon in September 2012. It was a wonderful feeling. No more fishy smells, no more bucking and rolling every time a ferry went down the river. The days of sun-tanning on deck were over. Our little startup was now taking a big step up in the world.

Our little convey arrived at Somerset House's service entrance. As we waited for the elevator to descend we saw a flurry of activity in the building, the source unknown. As we approached the ground floor, wheeling our loads into the corridor, we were we met by a horde of thin, beautiful, eclectically dressed women. Cameras whirred. Clipboards waved. Turns out the day we moved into our new offices was also the first day of London Fashion

Week. And Somerset House was the venue for the glamorous event.

I don't think I've ever seen so many stubbly engineers grin so broadly at one time. Now *that's* a welcome to our new office, I thought.

Somerset House remained our hub as we went through this revolutionary stage of Hailo's lifecycle. Throughout this stage we added numerous international cities and expanded the team quite quickly to 150 people, and kept growing until we broke through the 250-person barrier. Our focus was not only to grow, but also to ensure that our operations were becoming more profitable and that our business model was increasingly scalable.

Scaling

Everything up until now has been highly focused on experimentation and validation. It has been about using money wisely, about being as frugal as possible, to ensure you had the runway to get to real product–market fit – and then validate a business model that scales.

At this point all the rules change. You have reached an inflexion point.

Now is the time to pour fuel on the fire and *spend* as quickly as you can, to *grow* as quickly as you can. This is definitely the most exciting period of your company's life so far.

As we've seen throughout the book so far, success is about iteration and fine tuning. If you're well prepared – and have a bit of luck on your side – then scaling your business will be a case of injecting money into your growth and revenue engines while putting the right people and processes into place.

Scaling is not *just* growth. A business that scales is one that generates disproportionately more revenue for a given cost. Take a look at the top graph opposite – it represents a normal business. As it generates more revenue, the cost of generating that revenue increases in lockstep. That's good, but it's not great. There are no economies of scale – i.e. there are no operating benefits of being bigger.

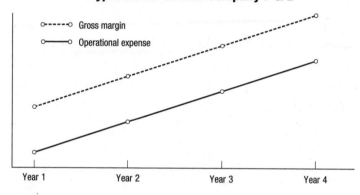

Hypothetical Normal Company P & L

The ideal scalable business model is represented by the graph[1] below.

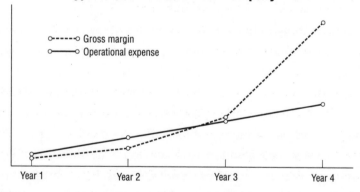

Hypothetical Scalable Tech Company P & L

Why is this curve more attractive? Well, over time, as your business gets bigger, your costs keep increasing – but your revenues grow *even* faster. That's the key of truly great businesses. That means there are big business benefits in being larger. These are called economies of scale.

It is possible to deliver this kind of growth – a great example is Google. Check out the graph below.

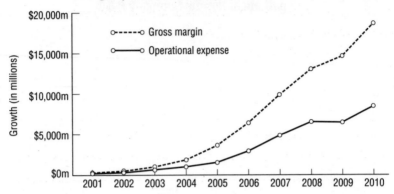

Google (NASDAQ–GOOG) Historical P & L, 2001–2010

It is clear – especially during the period from 2007 to 2009 – that costs can remain essentially flat and your revenues can grow substantially. In this case, Google's costs increased about 25 per cent but revenue increased by around 50 per cent. That's what you want.

Scaling requires doing *one* thing exceptionally well. Whether that is delivering a single excellent gaming app, focusing entirely on just processing payments or delivering the best possible taxi service, you have to make sure that you're doing something better than everybody else in the market. Don't try to do everything. Don't try to tackle multiple business problems. Focus on one core problem and deliver one great solution. Don't be under the illusion that you can solve several big problems simultaneously. It is only after you have demonstrated you have a scalable model – one that is firing on all cylinders – that you should think about expanding into other areas. Right now it is not about size: it's about sustainability.

The number of new people you will need to support this growth will vary hugely depending on the type of business model you are operating under.

Some apps – especially ones in the gaming and messaging sectors – are still small at this stage. WhatsApp, for example, hit a $1 billion valuation with 35–40 employees and Snapchat reached $3 billion with a team of just 35 people.[2] Other more operationally heavy apps such as Square and Uber are already going to have a couple of hundred people.

Why Scale?

A lot of the opportunities in technology – and especially in mobile – come with shorter lifespans than traditional businesses. Your business model could easily be in trouble if the market changes, if a competitor offers a more lucratively priced service, if the interests of users change or if either the iOS or Android platform introduces new features or disrupts existing ones.

The only protection you have against a fast-changing and competitive market is speed. You need to be able to take advantage of the opportunities that exist in the here and now, while constantly keeping an eye out for where new opportunities may lie in the future.

A great app that feeds the needs of a growing market attracts competition quickly, so you need to grab as much market share as you can before a competitor enters your space by scaling your business. Racing to the front buys you the advantage. So you'd better get ready for it!

You reach a clear tipping point when you're recognised as the market leader and at that point your competition become the also-rans. After this a snowball effect is set in motion, the market leader – provided they seize the opportunity – will command more attention from the media and their growing user base, and it is that loop that starts a powerful, self-reinforcing cycle. It's critical to grab this advantage.

Scaling is where you – the founder and entrepreneur – create the most value and generate financial rewards. This is, after all, the moment you have been waiting for – the time to show the world that your app can become the undisputed leader.

Beware Premature Scaling

It can be very tempting to scale too early. You will know as a founder that ego can be a big driver. It can be one of the reasons you try to scale prematurely. In businesses where revenues are low, or even in those where profit margins are not quite high enough, one easy-to-control barometer of success is headcount. Be careful – don't treat headcount as a vanity metric. It can prove very costly and can derail your company quickly.

Another common reason founders over-hire is to address a discontent team. When employees are not happy within a company – whether it's due to demanding workloads, fluctuating performance or weak leadership – it is easy to gloss over the issue by bringing in more people. Unfortunately, this tactic won't even act as a Band-Aid. It's a distraction and one that only delays a bigger pain – the pain of firing people down the line. Never hesitate to filter out the weak, underperforming employees. It sounds horribly Darwinian and harsh, but that's your job as founder and CEO. It is critical that you address any employee issues before scaling.

A third motivation to hire too early was one that hit us at Hailo. We thought that hiring was the solution to resolving a series of technology bottlenecks. A startup (and, frankly, every company) will always have a team or department that is a bottleneck for productivity. Thinking that simply adding more bodies will resolve your bottlenecks is not an effective way to think. In our case, we needed more engineers supporting the back-end platform. While hiring certainly helped alleviate one issue, it merely shuffled the bottleneck to another part of the organisation. The real root cause was simply a system that was too complex, one that should have been pared back in terms of features. If you adopt a liberal hiring culture, you're going to be adding too many people too early. On top of that, if you breed a reactive culture like this, the calibre of hires is likely going to be average. That waters down the culture, which has significant long-term impact.

You and your team need to know how to focus – and how to be lean. Focus on the root cause of the bottlenecks, by all means make hires, but don't sacrifice standards, and don't hire reactively. Hire because that's the absolute best thing to do.

It's a Different Skill Set

This is one of the areas that founders have a huge challenge with. The skills required for scaling a company are *very different* from the skills that are required to get a startup off the ground.

Getting a startup off the ground relies on very focused skills, especially in the art of product development and business development, or audience development. Scaling a business quickly takes nerve and it takes experience. It's less about reinventing the wheel, more about leveraging decades of best practice in management and operations, and doing so quickly.

So scaling a company potentially requires you as a founder to take a step back – and then very quickly learn a completely different set of skills. Rather than trying to oversee and implement the scaling process yourself, the best way to navigate through it is to bring senior people on board who have done it all before, to provide the skills and experience you lack.

Don't fret, though – this is your chance to delegate more operational work to someone else. With growth comes opportunity, and you should take advantage of the chance to put your business in the hands of people who know how to scale. That leaves more of your time to do the fun stuff.

These experienced people will be crucial in making sure your business scales quickly and effectively, so it's essential to make these hires wisely. I once received a piece of advice: 'Overinvest in your management team.' Why? An ineffective employee is bad for the company, but the impact of an *ineffective manager* reaches across the organisation and is multiplied by the number of people that person manages. You need to hire the best people you can get for management positions.

One of the most important processes we undertook at Hailo – even before our Series B funding was completed – was to identify the key hires that we would need to scale our organisation. This was a process that we knew would take months to complete effectively.

It's easy to *want* great people on board, but it's not always clear whom you need in place or what skills they should possess. We were lucky enough to draw on the experience of investors. Accel has helped scale companies such as Facebook, Supercell, Rovio, Kayak and Dropbox into multibillion-dollar successes. Sameer Gandhi, Hailo's partner at Accel, helped us craft a clear vision for what an optimal organisation would be like to take Hailo global, which helped us to recruit the right people into the right roles.

Chapter 29

Big Hitters

In many senses it's a relief to be at the stage when you can afford to bring in some big hitters. By now you already have a VP of marketing in place, driving your growth. You'll want to think about a more experienced VP of engineering – especially if your business model requires a big engineering team. If your business is more geared towards SaaS or enterprise, then you should be thinking about a VP of sales. If you're building a marketplace – and one that has significant operational complexity – a VP of operations will be an obvious next step. There is one role, however, that will be universally helpful no matter what kind of business you're focused on, and that's the chief operations officer.

Chief Operations Officer

One of the most challenging roles in any company is that of the chief operations officer (or COO). While the CEO is the mouthpiece of the company, charged with vision, product and seducing great talent to get on the bus, the COO is focused on the minutest of operational details, organisation and costs to make that bus run. Usually a bit more reserved, and a great listener, this person is key to making everything run like clockwork.

It is no surprise, then, that the best COOs usually hail from bigger companies, bringing a few more years of experience. This is an essential role to fill when a company demonstrates that it is growing fast and needs a master at the helm.

Coffee and cabs

At first glance, the worlds of coffee and cabs don't seem to have much in common. But what Starbucks has been able to do in terms of expanding a brand internationally, deliver amazing customer experience and a near-flawless operational model is nothing shy of amazing. And so it was for this reason that we targeted Tom Barr, a 13-year veteran of the Seattle-based coffee king, to manage the North American expansion and operations for Hailo.

Hailo was lucky to already have a seasoned operations guy, Caspar Woolley, as one of its cofounders. As we grew, however, we realised the task at hand was bigger and more complex than we thought. As we accelerated along the path of becoming a $500 million app, Tom Barr transitioned into the role of COO.

Tom is a quietly spoken, gentle giant. At Starbucks he worked across every division, understanding every detail of the company from product (he invented the Pumpkin latte), to operations, through to marketing and partnerships. He ended up managing a billion-dollar division of the company.

It was precisely because of this operational experience that we wanted him to help drive the expansion of Hailo. We needed his experience to help federate the disparate US taxi markets into a single consistent experience for users under a single, immaculately crafted brand, alongside a superb customer experience.

As you scale your company you need someone to inject a spirit of process and organisation – and one that scales to a huge level. This is not something that can be learned overnight by an inexperienced founder. It is best put in place by someone who has done it before. On many levels, scaling an organisation is not tied to a specific industry experience – but is more related to the type of organisation required to deliver a certain type of business model.

Everyone's Doing It

The addition of an experienced COO is a strategy repeated often by great technology startups. One of the most well-known success stories is that of Sheryl Sandberg, COO at Facebook. I talked about her earlier in the book and the 'cofounder dating' she engaged in with Facebook CEO Mark Zuckerberg. Before Facebook she spent seven years at Google, building up a huge team as vice president of global online sales and operations. She was also involved in launching Google's philanthropic arm Google.org.

When she joined Facebook in 2008, her immediate task was to help the company settle on a business model. Within a few months she led the decision to focus on advertising, and by 2010 had led the company to profitability. It's amazing what an impact a single individual can have.

Adopting a similar strategy, transportation app Uber hired Kees Koolen as their COO in June 2012.[1] Koolen was a seasoned operations guy, spending over 12 years growing booking.com to more than $9 billion in revenues and operations in 419,000 hotels in 193 countries.[2] Without a doubt he was instrumental in orchestrating the global rollout of Uber. It was that efficient growth that powered the extraordinary (and profitable) growth which led to their business increasing more than tenfold in valuation from 2012 to 2013.

Fresh off the back of a major funding round in 2013, Snapchat picked up a pretty seasoned COO, Emily White.[3] Previously she was the executive who was leading Facebook's Instagram advertising programme and before that she worked at Google.[4] It would be a safe assumption to think she might have been brought on board to think about how to scale the business and get on the path of revenue generation.

While it is very tough to land them, big hitters require a big salary and have high expectations. They can deliver game-changing opportunities to your company. That's probably why it is common in Silicon Valley for top startups to steal great people from each other. Francoise Brougher, former vice president of SMB (small and medium-sized businesses) global sales and operations at Google, joined Square as its business lead to help drive the company's growth. Bob Lee, chief technology officer at Square, also came from Google.

Chapter 30

Scaling Marketing

It is an exhilarating feeling to be able to invest more money in marketing. At Hailo we were able to finally get a number of great campaigns out of the door. We invested heavily in paid mobile marketing via numerous channels we knew were converting well, delivering users who loved smartphones, paying via credit card and, most importantly, taking taxis frequently.

We also opened the gates to our referral programme. By allowing both drivers and passengers to share codes offering £10, £20 and more off taxi rides for first-time users, we created a virtuous cycle as well as a lot of buzz around trying Hailo for the first time. Paying users started flooding in.

With your own Series B funding in the bank you will now have money to feed your own growth engine. In 'Step 3: The Hundred-Million-Dollar App', you put your VP of marketing in place, and bolstered the marketing team to further drive user acquisition and retention.

At this stage you need to put in place concrete goals around user acquisition, retention and revenues. As you grow each of these metrics, you will need to keep pushing down the average cost to acquire a user, while growing customer lifetime value. You – rather your VP of marketing – will need to make sure that you grow your marketing team so that each metric gets the attention it rightly deserves.

Delivering an increasingly profitable revenue stream at the end of this process is the massively challenging goal you're chasing – and that will require an increasingly large team.

Scaling the Marketing Team

As you trend towards becoming a hefty $500 million app, your VP of marketing will be the person charged with scaling your marketing team so that it can deliver. Let's have a look at what Hailo's marketing team looked like at this stage.

- The **VP of marketing** was leading the strategy to deliver global user-acquisition targets and global user-retention metrics, and shouldering shared responsibility with the city general managers to deliver on the global revenue and profit goals.
- There was a **senior marketing manager** who focused on managing the spend on user-acquisition channels, brainstorming and testing campaigns and optimising the customer acquisition costs, as well as monitoring the average revenue per user per channel.
- We hired a **driver marketing manager** – this began as a London-focused role and then expanded to support driver operations across all global cities.
- **Head of data** – this specialised person was hired after we raised Series B funding. Bringing a deep data background from Google, this person was responsible for helping structure our data to more easily extractable insights about how to optimise user-acquisition channels, retention rates and referral programmes – and help drive improvements in service levels.

- **Head of retention** – as with our new data specialist, we could afford to hire this person only after closing the Series B funding. Tasked with keeping his finger on the pulse of the retention rate – and increasing it – this employee is 100 per cent focused on increasing lifetime revenue from customers.
- We had a number of **business analysts** to support the marketing and product team (as well as the management team) to do analysis on all kinds of tasks related to improving revenues, optimising price points, and analysing the results of product tests.
- **CRM engineers** were dedicated to helping the marketing team run the customer-relationship management platforms we had in place. The main task was to ensure the system was being fed all the data about our customers and drivers in real time, so that the retention team could deliver the most targeted and relevant messages to users.

In addition to these core roles, which were based in London, we had a number of regional roles in our operational cities to better support local activities. These included:

- **Social-media managers** in all operational cities (and a head of social media in London).
- A **regional marketing manager**, who was in charge of North America, and another in Asia.
- **PR agencies** – relationships with public-relations agencies were managed centrally by our VP of marketing. At the same time, the general managers of our cities would help localise the strategies, while ensuring that a unified strategy was being followed using one central brief.

Our biggest leap at this stage was putting in place great people dedicated to the tasks of user acquisition, user retention and data. Once we were able to dedicate a person to each one of these areas, we saw huge improvements very quickly. Once again, it demonstrates that focus on the key metrics is critical – and you clearly need to have enough people to handle each effectively.

It is critical to hire people who want to get their hands dirty. Don't hire people who want to manage: there is always too much to do in a startup, no matter what the phase. Having people who can lead by doing is inspirational, and more cost effective.

Marketing Engineering Team

Ensuring that your growth engine is reliable requires a dedicated team – and doesn't just involve marketers. You need to remember that marketing is the engine that will perpetuate your existence; it will drive sustainable, reliable growth; and, if well managed, it will open all kinds of new horizons for your company. As a result, you need to make sure that it has the engineering support it needs to be truly effective.

Let's have a look at how the multibillion-dollar marketing goliath Groupon operates. It's quite relevant since the vast majority of its sales are now coming through mobile.

Rich Williams, SVP of marketing at Groupon, hails from Amazon. His approach is very customer-centric. He starts by defining very precise user segments, then collects copious behavioural data, tests numerous themes, messages and promotions, and then ensures every part of the process is tracked by powerful analytics. Everything he does is methodical; everything he does is measured. Putting a system like this in place – and running it – requires a team of engineers. When I talked through the process with him, he pointed out that it required a team of 40 dedicated engineers. Groupon does after all have more than 100 million users, and sends billions of marketing-related emails and push notifications.

Furthermore, the relationship between the marketing and

product teams at Groupon is very close. There is a tight-matrix reporting structure between the teams, where marketing and product requirements are morphed together and where analytics are pulled from a common system. In essence, marketing is focused on the channels and campaigns, and product is focused on the features and mechanics to deliver those.

Fostering this kind of symbiosis is a key to success, because, at a great company that loves its users, marketing is not so much about pushing information at users as making it easy and fun for users to pull the information that they want from your app.

Delving Deep into the Data

Up until now, off-the-shelf analytics solutions have been able to process the volume of users and activity that you've been seeing. As you become more successful, and your audience turns into millions and then tens of millions, you're going to start hitting the limitations of standard analytics solutions.

Solutions such as Google Analytics and Mixpanel can definitely handle well into the billions of analytics data points, but they will start charging you. Your costs will approach tens of thousands, and even hundreds of thousands of dollars per year. Again, not a bad problem to have.

At this point, you're going to hear the term 'big data' being thrown around. Apart from being a buzz phrase, it just means that your systems are now simply more capable of collecting and storing loads more data than was previously possible. For mobile apps, that means behavioural and transactional, but also geographical or positional data. If you record everything you can, that's going to add up quickly. On the one hand, that's super-exciting – clearly data is loaded with lots of potential insights about your users. On the other hand, it's pretty scary – it's complicated (and costly) to extract that insight from the mountains of data you're collecting. The good news is that it's becoming easier to extract more actionable insight from all this data.

As you spend more time digging into your data, one of the amazing results you will see is the detail you can observe about your users. It's like the first time you watch an HD Blu-ray movie. There is a massive amount of detail that grabs you – detail that you never before imagined possible.

This depth of information will give you a new set of dimensions across which you can measure and rank your users, enabling you to drive conversion.

You've already been grouping your users by customer lifecycle stage, by user-acquisition source and campaign. What you can now do is divide them into groups or 'customer segments' based on any piece of data that you're able to collect. One of the most interesting ones is their behaviour.

What do I mean by this?

Let's use an example to explain. At Hailo we put users in segments based on their behaviour. The first behaviour we focused on was how frequently a user booked a taxi. By digging into the data we could tell whether you were an infrequent taxi user (one or two times a year), an occasional taxi user (once a month), a frequent taxi user (at least once a week) or a power user (at least five times a week). We saw that the general population of taxi users tended to gravitate into these specific groups.

Why is this interesting?

For each one of these customer segments you can now calculate the annual revenue per user (ARPU). For an *infrequent* taxi user it was about £15 (assuming one ride per year times an average fare of £15). Since Hailo makes 10 per cent commission on each taxi ride (the rest goes to the driver), the gross margin stands at £1.50 per user.

On the other hand, a power user is worth £3,900 (£15 average fare times 5 times per week times 52 weeks) and Hailo pockets £390. Clearly it makes sense to communicate with power users differently, and so we devised very specific campaigns, rewards and even events to develop deeper relationships with our users. All thanks to data.

Partner Marketing

A new user-acquisition channel that typically opens up only when you're operating at scale is that of marketing partnerships. Once you have millions of app downloads, and millions of engaged users, people are interested in working with you (no surprises there).

If you can zero in on the mutually beneficial opportunities, there are massive rewards to reap. Well-executed partnerships can change the entire course of your company. And, since you've hired a brilliant VP of marketing, I am sure that they will be all over it.

Partnerships can be fun, and show why it's worth investing in your brand and your product and making users happy. At Hailo we experimented with a number of partnerships. It was quite amusing to think that most were initiated directly via partners contacting us through our website. The earliest opportunities came via American Express and MasterCard. These companies are quite hungry and keen to get involved in mobile, but the results were mixed.

We worked with MasterCard as part of its international 'Priceless' campaign. We were featured on a number of print and online ads, all over its website, and in a number of direct email campaigns – all featuring trackable discount codes giving users discounts on Hailo rides. Despite high expectations, and detailed tracking, we saw very little additional uplift to either app downloads or trips taken using the discount codes. We don't know why the partnership had limited success, but that's precisely why you need a lot of experimentation, a lot of trial and error in marketing.

On the other end of the spectrum, we had a very successful campaign with Coors, the beer brand. In a well-coordinated holiday promotion, across a number of Canadian and US cities, millions of dollars' worth of $10 Hailo discount codes, redeemable via the app, were distributed to Coors drinkers.[1] The promotion delivered a 'safe holiday' message encouraging revellers to take a taxi rather than drive under the influence. The campaign delivered tens of thousands of incremental downloads, and a massive volume of completed taxi rides. It was a great awareness campaign that introduced Hailo to a big new audience.

Square is a master at partnerships. In August 2012 it announced a deal with Starbucks, whereby the coffee giant would accept the Square Wallet app at all 7,000 of its locations (at the same time Square took a $25 million investment from the company).[2] The Square Wallet app stores your credit-card information, and then allows you to pay at any store accepting Square payments by simply scanning the Square App (in this case a QR code) on a Starbucks register. The key to success in this type of partnership is sustainable mutual benefit: Square clearly benefits from a high-profile partner, masses of PR, and great in-store positioning for its super-easy-to-use payments app; Starbucks gets a huge image boost as a progressive, technology-friendly company that already offers free wi-fi in all its stores, making it yet more attractive for the coffee-dependent tech crowd.

Game in Game

I opened the Clash of Clans app on my iPad, and before I could click through I was hit by an ad that popped up and took over my entire screen for the latest version of Angry Birds. In the early days of the Internet full-screen ads had largely been relegated to the domain of adult websites and free video-streaming sites. And yet today there is clearly a resurgence in the app world – because they are effective at driving downloads.

I was curious to see how common this practice was, so I launched the Candy Crush Saga app. Sure enough, I experienced the same thing, but this time the ad was for another King.com game, Pet Rescue Saga.

So what's happening here? In the most basic sense it's traffic – or, more specifically, download – bartering. One super-popular game does a deal with another super-popular game and they introduce each other's game to their respective user bases. Companies carefully work out whether one game's users are complementary or are going to cannibalise each other.

It turns out that many games are complementary, so it becomes a win–win situation, and often the gamer reacts positively because they are introduced to a new, fun game.

Naturally, King – the maker of Candy Crush Saga – wants to ensure that it's around for the long term. And that means de-risking its success from just a single game. And what's the best way to get downloads for your new game? Actively marketing to your own install base of 100 million daily active users.[3] It's a genius strategy.

While testing new user-acquisition channels is constantly required, you can't bet your business on it. As you pour money into effective marketing you need to be thinking about how to keep innovating your core product – and how to take it to the next level.

Chapter 31

Killer Product Expansion

When the best mobile-app companies scale, their products do not stand still. These companies constantly seek a new way to delight users *more*, become even more sticky or launch inventive ways to make it easier for users to spend more money.

At scale, the product challenge is slightly different. It's not just about launching 'simple' features. Real product innovation involves taking advantage of the scale you already have, of leveraging network effects to become even more powerful.

Network effects are massive: passengers use Uber because it has more cars than the competitors; drivers work for Uber because it has more passengers than the other apps. It's a system that feeds off itself, one that reinforces itself.

Merchants use the Square register app because it's simple, beautiful and gives them the ability to process credit cards. But now customers use the Square Wallet app to pay in a store because increasing numbers of merchants use the register app that enables Square Wallet payments. You can see where I am going with this.

Blurring Business Models

Flipboard has a very simple mission: to let people discover and share online content in beautiful, simple and meaningful ways. About 90 million people regularly use the app and it's one of my favourite apps, on both the iPad and the iPhone. The first wave of Flipboard's growth was fuelled by automatically creating 'personal magazines' directly from the social graph of its users. Flipboard pulls in stories from your friends' tweets, Facebook posts and Google+ accounts, and then cleverly curates them into a highly readable format. Its second wave of growth was giving users (including advertisers) the power to create – and distribute – their own magazines. Within months of launching the new magazine-publishing feature, users had created some 3.5 million of them.[1]

The stats are impressive, with the average user spending 15 minutes browsing the app per session. The combination of great user experience and a truly magazine-like interface has allowed Flipboard to charge rather lucrative advertising rates – '10 to 100 times what's normally done on banner ads', according to CEO, Mike McCue.[2]

So the app has already reached a very healthy revenue stage with a great user-centric service. But a great company isn't satisfied with its already impressive performance: it wants to disrupt things even further.

With a stable of thousands of existing advertising partners, and already high click-through rates on its current advertising (typically around 3 per cent, though some campaigns have seen up to 10 per cent), Flipboard started experimenting with e-commerce. In November 2013 it launched Flipboard Catalogs.[3] Now you can browse your favourite digital magazines on Flipboard, and purchase items directly featured in the stories. Think of it as the ultimate shopping catalogue. It already has the backing of numerous leading e-commerce brands such as Banana Republic and Fab.com.

The company – as it scales – is disrupting its own business model and making it more powerful. It has started to integrate three very important elements into a single business: content, community and commerce. When those three components come together they create a very strong ecosystem – one that is very hard to eliminate if it gets into the leading position.

Oddly enough, with the launch of Catalogs, Flipboard has actually improved the user experience by delivering on users' desire to easily and seamlessly purchase something from the magazine. Additionally, advertisers do not need to change their existing behaviour, and simultaneously have a new channel through which to sell their products in a highly engaging and social setting.

When a product feature feels like an organic fit or extension, there is a great chance it will succeed. This is how killer products scale.

A Square Family

We've talked a lot about Square so far. Let's turn our attention now specifically to its individual apps. After only a few years, it boasts a family of apps and services that work seamlessly together. First, it mastered mobile payments from a merchant perspective, and then sought to solidify a mobile transaction economy by launching a series of apps and services to fill in the gaps. The cadence of its product development has been impressive.

Square Register was Square's very first app, turning any iPhone or iPad into a fully equipped register ready to take credit cards and manage your business. Not only did it allow any small-business owner to accept credit-card payments, but it also enabled them to make informed business decisions based on transaction history. Based on the app's analytics, any small-business owner had the insight available to answer questions like: Should I open an hour earlier or should I stay open an hour later? Should I change the pricing on the iced mocha? If so, should I raise it or lower it? Do my sales increase when it rains? Do I sell more hot chocolate when it snows? Small businesses had never had access to such powerful information before.[4]

To address the customer side of the equation, Square launched its second app, Square Wallet, in 2011 (it bungled it a bit, launching it under the name Card Case, but recovered quickly). This allowed a user to store their credit-card information in an app, and then pay securely via the app at any merchant who was using Square Register. Luckily, there are about 2 million such merchants.

In 2013 Square added Square Market to the mix. This was an interesting development. Via Square Register, the company was already managing inventory for millions of merchants, so why not pop that inventory online – and give merchants another big channel to generate revenue? Market presents a great expansion channel in an already well-integrated ecosystem.

And, to top it off, in late 2013 Square launched its Square Cash app (and service), which enables users to send cash via email to any person, simply by cc'ing cash@square.com (and then linking their debit card securely to the transaction). It's a super-simple way to send cash to anyone – and the interesting element is that there's no fee. Clearly, this is meant to be a customer-acquisition channel to lure people into the Square ecosystem.

All in all, as great apps scale, they also continue their pace of innovation. Becoming bigger is not an excuse to slow down. If anything, you now have the means to build and test and improve all the things you couldn't before.

Product Extension

One of the most genius – and cool – product developments delivered by one of our billion-dollar apps is the move to the real, physical world.

Rovio, which had extended the Angry Birds brand into numerous areas, has been quoted as wanting to become the 'next Disney'. In 2012, it had inked more than 200 licensing relationships.[5] As a result, 30 per cent of its revenue was derived from franchising and licensing agreements that year. That was a total of $30 million in 2011 (it made $107 million in total for the year).

So, in 2012, Angry Birds hauled in $200 million in revenue.[6] The franchising and licensing business revenue was more than three times the revenue of 2011 and accounted for 45 per cent of Rovio's total revenue.[7] It is clear that this new revenue stream is very lucrative.

There is another clever benefit in pursing this strategy. Licensing – when done well – kicks off a very virtuous cycle. With more people walking around the streets sporting Angry Birds T-shirts, and children taking Angry Birds lunchboxes to school, and even Angry Birds stuffed toys appearing in children's bedrooms, there is a linkage to driving more downloads of the game. And, sure enough, that has been the case.

Reflect back to the early part of the book, when we talked about creating a great brand, name and visual identity. By setting itself up so well at the beginning, Rovio created a massively high-profit revenue channel, one that benefits every facet of its business.

A Vehicle for Every Occasion

One of the powerful aspects of being a mobile app is that you can provide highly local services. People can pop open your app when they are out and about. Providing a service that is very situationally relevant is part of the reason that apps have exploded in popularity.

At the same time, that presents a big challenge: for maximum effect you need to make sure that the service your app provides is aligned with what people want. In the world of games, people tend to play more or less the same ones around the world. Sure, there are customisations to make, but there are big populations of people who enjoy the same bird-slinging and candy-crushing action internationally.

But the big challenge comes if you're trying to provide, say, a transportation service internationally. Uber – the private-car app – has solved this elegantly. While its original business model solved the problem of San Franciscans not being able to find a taxi (there are very few in the city) by providing on-demand black cars, they found the problem was not the same in all cities.

So what did they do? Uber created a portfolio of transportation services that can be enabled in a pick-and-mix fashion on a city-by-city basis. It currently offers the following options around the world:

- UberLUX – luxury vehicles such as 7 series BMWs and S-Class Mercedes;
- UberExec – a midrange luxury vehicle, very nice, not crazy expensive;
- UberSUV – these are SUVs, as the name suggests, or Chelsea tractors;

- UberBlack – black cars are the normal private livery vehicles;
- UberTAXI – in some cities Uber also works with taxi drivers; and
- UberX – a service that allows normal people like you and me to become part-time taxi drivers without any of the regulatory hassle.

Uber cleverly tests each one of these services for product–market fit in each new city the company enters. Depending on the dynamics, politics and user demographics of a city, it can launch the right set of transportation options that are going to get the greatest usage.

In a city like London, where taxis are dominant and strong competition with the likes of Hailo exists, Uber focuses on the UberBlack and UberExec offering. In cities like San Francisco, where sharing a ride is all the rage (because there are not enough taxis or black cars) the UberX offering is growing rapidly.

Understanding how your product can achieve product–market – or, more specifically, product–city – fit in some businesses is key to becoming the undisputed leader in a space.

Chapter 32

Scaling Product Development and Engineering

After the $30 million Series B investment at Hailo, we wanted to get to grips with building out and industrialising our product and engineering teams.

Daniel Ek, Spotify's CEO, shared a great document about how his company scaled the agile software-development process to create an astounding music app, desktop app and website. It has a pretty geeky title – 'Scaling Agile @ Spotify with Tribes, Squads, Chapters & Guilds'. It's worth reading in full (I've included a link to it on mybilliondollarapp.com). The core takeaway is that, in order to remain agile and be able to quickly iterate and improve your product, you need to put in place a scalable architecture of teams (which Spotify calls squads) comprising product managers, scrum masters, engineers and testers, who own specific features or products (a squad might own the iOS app or the Android app or big parts of the back-end systems) and can independently release new features and bug fixes into production without depending on any other team (thus avoiding bottlenecks).

The next step – as your product and engineering teams grow –

is to ensure that these squads naturally group together into bigger groups called tribes. Within a mobile tribe, for example, you would have the iOS and Android squads. Similarly, different back-end squads would come together to form a bigger back-end tribe.

It's not typical for companies to think about this early on but, trust me, having a good idea in your head about how your team will scale from the very beginning is very useful (it helps you plan the big hires and get terminology and organisation in place early). As with any organisational change, it's a lot easier to get it in place when there are fewer people to convince, so do keep it in mind.

Hailo went through a long process of investigating organisational models that would work for our platform and marketplace business, and had numerous great conversations with the teams at Etsy, Skype, Spotify and SoundCloud. These conversations generated plenty of eye-opening moments, and led to the question of who we needed to put in place to implement this refined organisational structure that would underpin our ability to grow efficiently.

A Real VP of Engineering

While we attempted to impose a bit more structure on our growing engineering team, it proved tricky to do everything correctly. We realised that we needed external help. We knew we needed someone who had done this before. Thanks to our internal philosophy of 'always be recruiting', it was fortunate that, about this time, a man called Rory Devine came into our offices in Somerset House for a conversation.

Rory hailed from Betfair, a billion-dollar company and the world's largest sport betting exchange. He had joined the team early there and grew the engineering team from about 30 to more than 300 people within a few years. Immediately, Rory could empathise with our struggles: how we hadn't quite been able to achieve the efficiencies we were searching for; how the team structure or our speed of deploying new versions of the app weren't as fast as we knew was possible.

He quickly went to work walking us through his previous, similar experience. It was only a matter of weeks before everyone was on board with his streamlined strategy. We reorganised our development teams to focus on better-defined product areas; product managers were given a more narrow focus to allow them to deliver faster (and we hired more product managers); and scrum masters (also called 'agile coaches') were brought in to help planning, estimate workload and generally improve communication and process.

The change was truly amazing. Much of the shared frustration was removed as teams were empowered to more simply, and nimbly, get their code improvements into production themselves (and not be blocked by the systems operations team). We started to feel that we were firing on all cylinders.

One of the big changes we felt in the product team was the renewed feeling that we could focus more on actually developing new features and testing with users rather than just managing the backlog of work with the engineering team (great scrum masters made this possible).

From that point on, we felt more capable of building the engineering teams in the US as well as the UK. A new structure allowed us to create a clear separation between projects, allowed us to give teams the power to release their own software, and removed dependencies on other engineering teams that had been holding us up so unnecessarily before. It was a truly great time at the company.

Development that Scales

Whether or not you feel that you need a VP of engineering, or even need to build a big product development and engineering team, there are certainly a few core things to remember. Each of the engineering and product teams at our billion-dollar apps echoed the themes below.

- Focus on building a **product-centric organisation**. This means that, as you start organising people, organise them around the 'product chunks' that you're building – the app, the API layer, the admin and reporting systems, the back-end systems. Don't organise around technology or skills or functions – that only leads to confusion and doesn't build an organisation where every single person is focused on building the greatest product experience for your users.
- Make sure that you build teams into **self-empowered squads or units** ('cells', as Supercell likes to call them) that can deliver new features and bug fixes as an independent unit, without creating dependencies on other teams. That means creating teams with dedicated product managers, designers, engineers and testers all working together.
- Invest in **great scrum masters**. They allow the agile software development process to scale neatly; they empower product managers to focus more on developing features, less on process. Exceptional scrum masters improve communications between all teams, and remove dependencies. They make development teams happier.
- **Invest in people, not infrastructure**. Hold out for exceptional engineers, the 10x employees, the ones who deliver above and beyond. Snapchat managed to build a platform that sends 400 million snaps per day[1] with a team of 35 people; WhatsApp handled 18 billion messages on New Year's Eve 2013 with a team of about 55 people.[2]
- If you are planning to go cross-platform (supporting iOS and Android, as well as other mobile platforms like Windows) and scale rapidly, it's imperative that you **focus on the API first**. You want to ensure that you build what's known as a service-oriented architecture.

If you keep these things in mind, you're not going to run into any major problems. Stay abreast with best practices in software engineering and you'll be able to see the benefits of tuning the organisation of your engineering team. Maintaining a harmonious environment that allows great engineers to focus on developing amazing solutions and products is key. Given how tough it is to find, and retain, great engineers, you want to make sure that, once you've persuaded them to join you, they're going to stay for the long run.

Cost of Great Engineers

The war to seduce the best engineers is raging. It is perhaps most acute in Silicon Valley, but it is definitely being felt worldwide as more startups vie for a small group of exceptionally talented individuals.

You already know that you're going to have to build an engineering team to scale your business. You'll need engineers to build a better app, build better and faster databases, process mountains of complex data and support the marketing and operations team with data and analytics. You're going to need all kinds of different engineers on the team and, if they're any good, well, you're going to have to steal them away from the competition. So let's have a look at what software engineers are earning these days, and what you're going to need to make it interesting for a top one to come and work on your team.

First, let's talk base salaries. For a good engineer we're talking between $100,000 and $130,000. That doesn't include perks and bonuses. That number is based on data from the top 25 tech companies in the world.[3] Glassdoor, a career website, culled data from 3 million salary reports, company reviews and interviews from employees at more than 210,000 companies, and found Google had the highest average base salary for software engineers at $128,336. Facebook came in at number two with a base salary of $123,626.

So that's a killer combo: the most exciting places to work as a software developer also pay the most. Where does that leave the rest of us? How are startups supposed to compete with numbers like that?

What about the *average* software developer? In the US, according to Bureau of Labor statistics, the average salary for a software developer is $90,530.[4] Glassdoor's own figure suggests a comparable $92,648. So, while it's less than the top guys get, even an average software developer is a pretty hot commodity.

Company	Average base salary
National average	$92,648
Google	$128,336
Facebook	$123,626
Apple	$114,413
eBay	$108,809
Zynga	$105,568
Microsoft	$104,362
Intuit	$103,284
Amazon	$103,070
Oracle	$102,204
Cisco	$101,909
Yahoo!	$100,122
Qualcomm	$98,964
Hewlett-Packard	$95,567
Intel	$92,194
IBM	$89,390

(Average base salaries based on at least approximately 20 software engineer salary reports per company of 2012 (08/10/11–07/10/12). Data taken from Glassdoor Report: 'Software Engineer Base Salary Comparison'.)

Price of Design

In recent years it has also become increasingly competitive and costly to seduce great designers to join your team. That makes sense especially as the world of mobile matures, and where delivering a great user experience becomes more important. In the US, the average software designer salary, at the time of writing, is around $77,490.[5]

In some cases the situation can become even more competitive. Some tech companies are actively pursuing entire design teams or design agencies as the competition for talent heats up. Over the last few years Facebook has acquired three separate design companies, both in the US and elsewhere.[6] In 2006 it acquired an entire team in Amsterdam (called Sofa) to lead a design overhaul,[7] and a design research firm called Bolt Peters in 2012 to work improving user experience.[8]

Square, the mobile payments app, has also been forced to acquire entire design firms to bulk out talent during its scaling phase. In 2012 it acquired the firm 80/20 to work specifically on app-interaction design.[9]

Chapter 33

Scaling People

It's easy to talk about finding great people and hiring them. It's a lot tougher in reality, when you have to balance so many other pressures when growing a company quickly and effectively. One of the biggest advantages you're going to have in being able to scale your team to support your growth is that big hitter. Having experienced operational leaders in place will completely change the tone within your company, eventually leading to much more calm and focus. No longer will people be fretting about the day-to-day issues, the processes. They will be back to focusing on how to build something great to make users happy. Ensuring that you have key leaders across the team – from operations, to marketing, to engineering – will make sure that every part of the company benefits from that additional bit of experience and eventual calm.

Irrespective of great leadership being in place, there are other organisational challenges that emerge. There is a magic point around 150 people when an organisation changes. According to British anthropologist Robin Dunbar, 150 is the limit to how many stable relationships an average human being can maintain. That number is now indelibly linked with his name – it's called the Dunbar number.

After that point it's increasingly hard to remember everyone's name, and an organisation becomes more bureaucratic and political as teams are forced to break up. Problems that were easily solved by yelling a question across the room now require emails to be sent, meetings to be called or adventures down the corridor or up to the next floor.

As a result, this section will focus primarily on the challenges faced by startups when they grow from a small team to a company of over 100 employees. The challenges do mount quite quickly. At Hailo we were an international operation from the start. While initially it was relatively simple to keep all the product and engineering in one place, and provide all our cities with regular updates every couple of weeks, it was a different case for the marketing and operations teams.

And then there is the flipside: having too many meetings and not enough time spent actually doing things.

The key to efficiently scaling your team is not only getting new people to join, but also making sure you keep the existing (and rapidly growing) team of employees exceedingly happy. This involves a lot of process – but it shouldn't lose the human touch.

Founders Can't Do It All

It's not just the team members who feel this change as the company grows. The people who feel it the most are certainly the cofounders. And there are a great many examples where, despite wanting to scale and grow the business, they find it incredibly hard to yield control and delegate. There are great stories of technology company founders who still screen the CV of every single new hire at their companies. In some cases they even insist on *interviewing* every single candidate personally.

In Google's early days, cofounder Sergey Brin would interview every candidate[1] and would know right away if he was interested in hiring them. If he wasn't interested, he said, 'I would try to spend the next hour trying to learn at least one thing from the person so that the meeting wasn't a waste for me.' How very selfless of him!

I think there are many virtuous things about this – and I think it evokes a passion and commitment to building the right kind of company. But there clearly comes a point where the process can't be sustained – it becomes a bottleneck.

At the same time the opposite – a founder or CEO who is not at all involved in the team-building process – certainly doesn't work, either. While I personally think the hands-on approach is correct for the beginning phases – by the $100 million stage and moving into the next stage it becomes impossible to sustain.

A great solution is one that Amazon has put in place. Early on at Amazon, Jeff Bezos, the CEO, handpicked a team of people he thought particularly understood the culture of the company, and whom he trusted to make independent hiring decisions. For all future hires, no matter for what team, the final say would be made by a senior executive with this 'power of hire'. This ensured that there was an incredibly high – and consistent – bar for hiring. And it also insured against managers making rush hires for their individual teams (rather than holding out for the best candidate).

Scalable Recruiting

At Hailo it took us a while to find the ideal person to head up talent acquisition. It's a very tough role to hire for in a fast-paced technology startup. We were always recruiting and interviewing people for any number of open roles. That meant a lot of logistics,

schedule management and then a lot of contract-related documentation and onboarding. In short, it was a lot of work.

In the spirit of operating lean, we came up with a rather novel
solution. We put in place a very experienced and high-energy executive support team. We recruited a super-ambitious executive
assistant who worked with the CEO and the management team
who basically managed every type of administrative task – including everything to do with recruiting.

When you hire someone experienced and hungry in this role,
you get the best of all worlds. You effectively have an intelligent
person helping coordinate all kinds of activities as and when necessary. You could call this role an 'office manager', but we found
better success marketing it as an executive assistant – and then
offering a healthy salary as well. You definitely do get a much
higher calibre of person here when you dedicate a good budget to
it.

Within three years this executive team grew to four people and
helped us get to the 200-employee mark – so it works. It was at
that point we needed to get a chief people officer in place.

Growing the Team at Square

Square is the gold standard of execution. It has a clear vision to be
not only the best possible mobile cash register and mobile card-
payment processor, but also the best mobile-app wallet – and now
even a marketplace.

It has reinvented the modern mobile commerce ecosystem in a
number of ways. With so many moving parts it needs a big team.
Its business model, which is predicated on tiny margins, needs
huge transaction volumes to be profitable. So scale is basically at
the centre of everything it does. Accordingly, it has needed to raise
a lot of cash (more than $340 million) to realise this ambitious

vision. Square started in January 2011 with 37 people; by January 2012 it had 210, and ended 2013 with about 600.

In an interview[2] the former COO of Square – Keith Rabois – shared three key things that helped Square grow its team:

> First, we have a very clear vision of what we are trying to do, and it's very ambitious, and that attracts the best people in the world. People who are really good at what they do, elite engineers, elite designers, are all motivated by difficult challenges and confronting them.

> The second thing is that we have significant and substantial momentum in the marketplace, so it reinforces that vision.

> Then the third is that we have a positive impact. We have a mission of helping local businesses thrive so that they can hire people, so that they can help the U.S. economy.

But the ambition doesn't stop there: there are bigger plans to hire even more amazing people.

The Road Will be Bumpy

It's worth mentioning that even at the best companies it's not always smooth sailing. Building teams is complex because people are complex. Hiring experienced people, and especially big hitters, comes with politics and power plays.

I don't want to pick on one company in particular, but in the case of Square, there have been a lot of challenges. Delving into why is not something that's easy to achieve, and that's not my goal. My goal is to communicate that, just because a senior hire leaves, it doesn't mean the company is doomed. Alyssa Cutright, who was Square's vice president of international, lasted barely a year on the job. Alex Petrov, another senior hire who accepted the role of vice president of partnerships, never actually started working at Square.[3]

The biggest – and most trying – departure was that of Keith Rabois. The details of his departure are not important (but, in case you're interested, the Silicon Valley gossip train is only a Google search away). Even in this case, where the number two of a company leaves, that doesn't (necessarily) spell disaster. The whole point of having an experienced management team and a robust organisational structure is that the company is not dependent on any single employee, and will endure through tough people complexities.

In any high-growth startup there *will be* drama and some amount of employee churn at every level. My nugget of advice: expect it to happen; deal with it; move on.

Common People Mistakes

So what leads to a lot of these problems?

Companies all make similar mistakes when it comes to people management during high-growth situations. Here are some of the big ones.

HIRING TOO QUICKLY. Make sure you hire well – and not in a hurry. Ensuring that there is clear culture fit, great technical skills and a true personality fit within the team is worth the extra time. Everyone in the company should be involved in recruiting and getting to know people through the recruiting process. You're in the business of building a company, not just an app, and if you hire people to last then they will help build something that will last.

FIRING TOO SLOWLY. There's a simple test to see whether you should fire someone. Ask all the people with whom they regularly work and ask how things are going. If you start getting some funny responses, then you know there's an issue. If you figure out there *is* a big issue, something you didn't pick up in the interview process, or a fundamental disagreement, then try to fix it. If that doesn't work, do yourself a favour and cut them loose. The startup environment is demanding enough as it is. Don't make it harder. And make sure you hire everyone with a 90-day probation period, during which either party can call it off with a week's notice.

FORGETTING TO KEEP PROMISES. In a technology company, where a lot of the team are engineers, it's too easy to forget that they are not the most talkative people – and won't necessarily tell you when you have forgotten to deliver on something that you have promised. Generally, the omission is not intended, but when people think they're being ignored they will start looking. Quarterly reviews ensure you don't forget promises – especially ones about personal development.

HAVING NO ONBOARDING PROGRAMME. Make sure that you put together at least a basic onboarding or induction programme. It gives new team members a great first impression, whereas lacking one entirely is particularly unimpressive (especially at the stage of 50+ people). This not only helps the company organise itself, but it encourages managers to be particularly organised and attentive.

Startup Cribs

I used to think that having a spartan office, populated with cheap office chairs and sporting desks made of cinderblocks and recycled doors for tops, was the way to go. Well, that's both right and

wrong. During the early days it's all about finding an OK space –
one that won't cost the earth but that has a reliable, and blister-
ingly fast, Internet connection. With those basics in place, more or
less anything else can be forgiven.

But, when it makes financial sense, offices become one of the
key ways that you make your team happy and motivated to come
in every day to make your vision a success. Giving your employees an office environment that is comfortable, fun and something
they can call their own is a huge way to give them yet another way
to see that this is their company just as much as yours.

The scaling process can be particularly hard, and anything you
can do to smooth out that ride is going to be a good thing.

When Hailo outgrew the HMS *President* floating office, the
entire company was involved in the process of finding a new location. Everyone was encouraged to suggest areas, search for
buildings and generally help out in the process. It was through a
friend of a friend that we were informed that one of the most beautiful and historic sites on the banks of the Thames – Somerset
House – actually also housed a number of small companies and
startups.

As we saw in Chapter 28, it was close by, which was a big plus.
The building had huge charm, restaurants and art galleries and
was a popular backdrop for all kinds of events (not least of all
London Fashion Week). Possibly the best feature was its enormous
rear veranda, spanning the entire back width of the building,
which, I might add, also had a bar. It didn't take much effort to get
people to hang out for a drink after work.

We invested a good amount of money moving into a newly
refurbished wing of the building. We took time to bring in architects
to create a warm, fun environment that reflected our personality. And, once again, we called on personal relationships to get
the work done at preferential rates: a good friend called Nils
Fischer, who is a leading architect with Zaha Hadid, was able to

transform the space into something that increased productivity markedly.

While the final space was still not something that could compete with the ridiculousness of Google or other Silicon Valley royalty, it was one that outstripped just about all startup space in London, and certainly evoked the Hailo personality. Moving into that new environment was an exciting step forward – and upward – in the journey of becoming a billion-dollar app.

Chapter 34

Scaling Process

Even more important than having a lovely office is making sure that, as your team grows, things run smoothly. It sounds pretty boring on the surface, but frankly, when put in place, it is one of the most satisfying stages of building a company.

Typically, you can just force your way through various goals and challenges without too many problems until you're up to about 50 people. Process might be lacking, goals might be a bit vague, but momentum – so long as things are going well – will get you through it. But, at the first signs of plateauing performance, stress can creep in. The good news is that a bit of process can get things back on track quickly.

So, irrespective of your size, I suggest having a number of simple processes in place to be in a position where your company can weather any kind of hiccups – and continue to scale quickly.

Two-Pizza Teams and Internal APIs

Amazon's business model is predicated on the thinnest of margins, and as a result the company has been able to innovate in amazing ways that don't require massive capital investment. The principles therefore apply to startups looking to stay lean. Amazon is undoubtedly one of the most entrepreneurial companies in history. Two reasons Amazon is able to keep its momentum even with 97,000 employees are pizza teams and APIs.[1]

TWO-PIZZA TEAMS. Jeff Bezos structured Amazon as a decentralised company where small groups can innovate independently and are free from the inherent problems of groupthink. He introduced the principle of the two-pizza team. If two pizzas can't feed a team, then the team is too large. That limits a task force to five to seven people, depending on their appetites.

INTERNAL APIS. We talked about application programming interfaces (APIs) in Chapter 2. Think of them as clearly documented instructions about how two pieces of software should talk to one another and exchange data. At Amazon, Bezos mandated that every internal product or feature should have an API. This means that it can be more easily shared (and used) by internal product and development teams. But it also means that the option exists to allow external developers to use it as well.

The added bonus is that APIs lay the groundwork for the productisation of internal tools.

That simple declaration created an IT, as well as cultural, architecture that catalysed the growth of Amazon Web Services. Within a few short years of its launch in 2006, the service was already a billion-dollar business.[2]

In short, small teams can run fast and innovate because of their size and the fact that they're not reliant on the technology from other teams.

Move Fast and Break Things

Facebook created a culture of agility, promoting a philosophy to 'move fast and break things'.[3] Mark Zuckerberg explained the company's 'hacker way' in a letter to investors:[4]

> Hackers try to build the best services over the long term by quickly releasing and learning from smaller iterations rather than trying to get everything right all at once ... We have the words 'Done is better than perfect' painted on our walls to remind ourselves to always keep shipping.

Prioritising speed over perfection is a powerful way to maintain momentum as you scale, particularly as your organisation becomes more complex, and bureaucracy invariably creeps in.

Makers and Managers

What a lot of startups don't realise – and therefore fail to take into account when they are implementing processes – is that tech companies fundamentally have two types of people: managers and makers. It's very important to draw the distinction now, as it impacts on quite a few of the core processes you'll be adopting as your company grows.

If you're an engineer at heart, this will make sense instinctively; if you're from the business side of things, pay especially close attention.

So what's the difference between makers and managers? A maker is anyone who is involved in a creative process, a person who works best given time to think, read, invent, discover. This person needs time to get in the zone, and requires longer periods of time to deliver value. Once makers are given clear goals, it's best to let them sort out their own process to deliver what they need to. Engineers, designers and creatives are all makers. The manager is someone who is a boss. They have a day that is sliced into one-hour slots – and it's pretty much a collection of meetings. With so many things to be on top of, the default is to give everything priority for one hour – and see how much gets done.

Why is it important to draw the distinction?

Makers don't function efficiently if they are constantly being disrupted – or if they're being called into meeting after meeting. In fact, they hate meetings. If they're in a meeting, they're not making.[5]

I am not saying that makers should never attend meetings. As we've seen, meetings first thing in the morning – daily stand-ups, where developers and designers quickly discuss what they did yesterday, what they're doing today and anything that's blocking their progress – are helpful.

Facebook has a loose policy called 'No-Meeting Wednesday', when engineers are encouraged to 'stay focused on building products'. Y Combinator's Paul Graham echoes this idea on his blog:

> When you're operating on the maker's schedule, meetings are a disaster because they disrupt the creative flow. You can't write or program well in units of an hour. That's barely enough time to get started.[6]

For a maker, a single meeting can destroy their flow for a whole afternoon, by breaking it into two pieces each too small to do anything meaningful in.

The best approach for organising meetings on a maker's schedule is to have them at predictable times of day. There's a reason that agile software development encourages short meetings at the very beginning of the day: so the rest of the day is free to write great software!

A Venture Capitalist's Perspective

There are exceptional cases when you see both the maker and manager personality in a single person. Fred Wilson, a very geeky partner at the venture-capital firm Union Square Ventures (and also an MIT grad), has helped many blockbuster Internet startups and apps scale their processes into robust billion-dollar companies.

He brings a rich perspective to the table – one that leverages a deep knowledge of how complex engineering systems work and how that can apply to growing technology companies.

'I think the work you do on computer systems can be a metaphor for the work you will find yourself doing on people systems,' he says.

Wilson highlights the following characteristics of scaling people processes.[7] It's fascinating because we experienced them all during the evolution of Hailo:

- Things that work well at small-scale tend to break at large-scale. You need different people, processes and systems as a company grows.
- You need to instrument your system so you can see when things are reaching the breaking point well before they break. You need to

implement employee-feedback systems, ideally real-time systems, so you can measure how a team is functioning over time. This point is very easy to overlook — especially if you're growing rapidly. We'll look later at how to put in a lightweight structure around employee reviews.

- There is always one problematic component in a system that causes the majority of the scaling problems and must be rewritten. Team members, particularly super-talented ones, who cause friction and pain in the organisation need to be let go, no matter what the cost. It's always tough to fire team members who perform exceptionally well but generate issues with other team members. The health of the organisation is always more important than a single individual. We had to make the call to remove key team members a couple of times at Hailo.

- Loose coupling of components is critical. You can't have one component fail and take down the entire system. Build resilience into your organisation, processes and systems. By hiring the best people — and ones with a variety of strong talents — you're going to build in redundancy, and the ability to weather big team challenges, as Square was able to.

- Blameless postmortems are the key to learning from a tech-ops crisis. Fear-driven organisations do not scale. Calmness in chaotic situations is another trait learned over time. Blame is never helpful when working with others — and certainly not in public forums. While you should always keep track of how employees are performing, fear and blame are not strategies that build a great organisation.

- Overreacting to a crisis is likely to make it worse. Remaining calm in the face of adversity is one of the signature traits of great organisational leaders. This is essential for the founders, the CEO and the COO. Again, this is something that comes with experience. Every startup I've ever known or worked with has experienced near-death experiences. It's the tough who survive.

- Overbuilt systems are hard to implement, manage and scale. Build the organisation you need when you need it, not well in advance of when you need it. This is a question of fine balance and one of the toughest things to accomplish for a startup. Timing is everything, as is investing the right amount of time and resource *at* the right time to make sure you have the right organisational structure in place.

Scaling Decision Making

Putting in a systematic way of making decisions – from big to small – and then communicating them effectively across your company is something you'll need to polish throughout this stage. You won't be able to grow rapidly (at least not in a smooth way) until you have a number of things squared away.

Decisions are at the heart of any organisation. *Fast* decisions are at the core of every *great* organisation. As the founder and CEO of your company, you have the responsibility to come up with a crisp vision for the future of your app. One of the first things that you need to get in place is a way to set annual and quarterly goals for your company.[8] That's something you will need to do in close consultation with your senior management team.

ANNUAL OFFSITE MEETING. You need to ensure that you're on the same page as your senior management team. Taking time out at least once a year to discuss nothing but vision, strategy and how to get there is crucial. It's important that you have a bigger vision – and goals – for where you want to go. And it's even more important that you all agree on that vision. Given how critical this is, it should not be derailed or interrupted by anything – so take a few days and do it offsite.

QUARTERLY MEETINGS. It's critical that your senior management team maintain a balanced view of both the tactical and strategic priorities of the business. If you're using objectives and key results (or OKRs, which I talk about below), this is the perfect forum to evaluate whether the key quarterly objectives have been met for every department and team. It's the time to figure out why objectives were missed, and also the time to map out – and agree – the top-level OKRs for each department for the coming quarter.

You can also take this chance to prepare for your board meetings, which – now that you have professionals on board – should typically be happening once every three months.

MONTHLY STRATEGIC MEETINGS. Assuming that you've implemented OKRs in your organisation, this will be a meeting with two goals. First, it should see how the company is tracking against its current objectives; second, it serves as a forum to adjust the strategy on any level should it be necessary. You want to always walk a fine line between executing on a clear strategy – as well as remaining nimble enough to make continual course corrections.

WEEKLY TACTICAL MEETINGS. It's a good idea to hold these on Monday mornings, since it gives you the CEO a clear view on everything that is going on this coming week. A quick check-in on a Friday afternoon doesn't hurt, either, to signal that everything is on track or to highlight anything problematic. I've seen it executed with the senior management team, or a smaller group, such as the CEO and the COO. This ensures that all key players in the team are the on same page every week.

Find a meeting framework that works for you and your team. Be religious about putting the critical meetings in the diary well in advance – I am talking months here – so that everyone knows when the key dates are.

Devoting a day or two per quarter to this is very important – something we at Hailo unfortunately learned the hard way. In companies with several international offices it is particularly important to schedule regular face-to-face sessions, because, in a frantic startup environment, a lot can be 'lost in translation' in emails, and, similarly, things can be instantly clarified when everyone is together in person.

And then all of these are supported by the slightly looser structure of daily check-ins. More informal meetings should ideally be kept to a minimum, and ideally pushed into the slots for the more regular meetings. I've also outlined below some of the best ways to keep meetings meaningful and relevant.

This is a very basic meeting structure that can be mirrored at every level – from the senior management level to the departments and teams. It scales nicely with any size of organisation.

Meaningful Meetings

As well as putting an effective framework in place for meetings, you need to make sure that each individual meeting is as meaningful and productive as it can possibly be. A well-run meeting is a great thing: it empowers people to make decisions, solve problems and share information. Badly run meetings, on the other hand, not only waste precious time, but tend to frustrate and demoralise people.

Google – one of the most successful technology companies of all time – has spent a lot of time trying to better understand how to run effective meetings. Below are some of Google's recommendations about how to make meetings more effective.[9]

- Every decision-oriented meeting should have a clear decision maker, and, if it doesn't have, the meeting shouldn't happen.

- Meetings should ideally consist of no more than 10 people, and everyone who attends should provide input. If someone doesn't provide input, they probably shouldn't be there.
- Being punctual is important.
- Preparing for meetings is key. Every meeting should have an outline, agenda and goal. All this should be communicated upfront, so that people can come prepared. If the material is not prepared, the meeting shouldn't be happening.
- A meeting is about sharing information and having a discussion with a clear objective in mind. A meeting without an objective should not happen.

After Larry Page took over as Google's CEO, he limited most meetings to 10 attendees and pushed managers to outline clear goals before scheduling a gathering.

He also talks about how rapid decision making is integral to becoming and staying a great company. There are simply no technology companies who can afford to take a long time to make decisions. The competitive advantage is making decisions rapidly, acting on them and then making course corrections down the path. Decisions should never wait for a meeting. If it's critical that a meeting take place before a decision is made, then that meeting should happen right away.[10]

Chapter 35

Financing at a Big Valuation

A t this altitude, everything becomes a bit blurry. Your app is worth hundreds of millions of dollars. If you've executed flawlessly, you might be worth more than that. You're also a bit of an old hat at raising money now, having talked to myriad investors over the years it took you to get to this stage.

Just like most other aspects of your company at this stage, raising finance is going to be a different process to before. It is going to be based increasingly on current business performance and financials, rather than on future potential. You will also most likely be talking to another group of investors, ones who specialise in investing larger amounts of money, who are generally more conservative and as a result are going to be digging into your financials.

At these mature stages only the biggest VC firms such as Accel, Index, USV, Sequoia, KPCB, Atomico, Greylock and Andreessen Horowitz – and potentially private equity or hedge-fund investors – are going to be able to come to the table. That's because companies at this stage are typically looking to raise a large chunk of money – in the tens of millions of dollars – and there is only a small population of investors with pockets deep enough to play at this level.

At this stage, investors have more conservative assumptions about the returns they can generate. They know that their investment is not going to increase in value 10 times over – they're going to be quite happy to with a 2x, 3x or 4x return.

Money to Grow Frantically

Uber is a prime example of an app that has managed to make venture capital work. In August 2013 it announced a long-awaited Series C funding round. It managed to raise a whopping $258 million dollars from investors including Google Ventures, at a $3.5 billion valuation.[1] That's an incredible amount of cash to raise, but, more importantly, the company had to give away only around 7.5 per cent of its equity for that.

What drove that incredible Series C was the fact that the company was able to increase its valuation 10x in a period of 18 months. When Uber closed its Series B financing it had a valuation of $330 million.[2] That's an extraordinary achievement for any company, and it's largely the dogged persistence of CEO Travis Kalanick that's driven it.

The company processed just shy of $1 billion in fares in 2013, delivering top-line revenue of around $200 million. By the beginning of 2014 it was operating in over 60 cities and 26 countries.

So, even though the valuation seems high at first, once you dig into the underlying numbers it starts to become a lot more justifiable. More importantly, Uber's investors appreciate that, in order to stay in the lead, the company needs to capture – and cement – its lead in as many cities and countries as possible as droves of local competitors appear.

When scaling is executed well – namely off the back of a great product–market fit, a scalable business model and a solid growth engine – massive valuation creation is certainly possible in a very short time.

Billion Dollar Investors

It's always good to know which investors come with the Midas touch. I've counted 49 venture-capital-backed private companies that have raised a funding round at a valuation of $1 billion or above.[3]

Sequoia Capital has backed 11 of the companies, including 4 at the Series A funding stage or earlier. Andreessen Horowitz, Digital Sky Technologies, Goldman Sachs and New Enterprise Associates have invested in seven billion-dollar startups.

Investor	Number of Billion-Dollar Startups Invested In[4]
Sequoia Capital	11
Goldman Sachs	7
New Enterprise Associates	7
Andreessen Horowitz	7
Digital Sky Technologies	7
Khosla Ventures	6
Accel Partners	5
Tiger Global Management	5
T. Rowe Price	5
Founders Fund	5
JP Morgan Chase & Co.	5
Kleiner Perkins Caufield Byers	5

Summary

When we approached the $500 million checkpoint at Hailo it was crazy to look back and see how much so many things had changed. A seat-of-the-pants operation transformed itself into a lean, mean operational machine. We had teams in 13 cities and 6 countries. Our marketing team was based in London, but, due to a highly efficient structure, was firing efficiently in all our markets.

The engineering and product teams had matured into a finely honed machine, regularly rolling out feature improvements around the world and testing changes, tweaks and improvements in a continual fashion with a clever, homegrown app-testing framework. It was a far cry from the handful of guys hacking an app together on the HMS *President*.

You'll definitely have a big team – probably anywhere up to a 100 or so people (again this will vary wildly depending on your business model). You will have strengthened your senior management team, having brought in people who have done it before. With adult supervision you'll have more confidence that you've put in place the right systems and processes.

At the same time as getting experienced leaders in place, you have maintained your close communication with users – and are understanding what they want. Your team is spending 80 per cent of its time improving the product and 20 per cent on blue-sky thinking and innovation.

Your business is growing wildly because, along with your VP of marketing, you have put in place a robust user-acquisition strategy, and you're continually finding new channels to experiment with. You have demonstrated that you do understand what your customer acquisition cost is, and the marketing team is driving it down quarter by quarter. You're focused on new initiatives to increase the average transaction value, which is driving up your annual revenue per user and lifetime value.

You're tracking all your key metrics, and using data to further segment your users and optimise how to deliver the best experience to all of them. This has allowed you to focus on profitability. You've delivered an amazing app, which has now become an amazing company.

But the job is never finished, and there is always some new issue to solve and challenge to face. There's always a new competitor popping up; there are always people and relationships to manage.

What is most exciting, however, is that you are on the cusp of breaking into the Billion-Dollar Club. You're at the edge of a cliff that few have ever ventured over.

STEP 5

The Billion-Dollar App

The Promised Land

Going from a Five-Hundred-Million-Dollar App to a Billion-Dollar App

App

- It's about keeping the momentum going. While driving constant improvement of the app and driving all metrics in the right direction, you need to create a crystal-clear vision for your product and take systematic steps to realise it.

Team

- Since you have all the key management in place, this now becomes a process of building the best scalable organisation to support the growth of your app.

- Different businesses will require different people structures, and by bringing in experienced people you need to make sure that you have the right structure in place as the foundation for your business.

Users

- Above all, the strategy is to keep your users ecstatically happy; if you maintain that, everything else will flow.

- Along with systematic acquisition and retention, you need to make sure that you have great communication channels with users to make sure the app is evolving with their needs and desires.

Business model

- Your business model is going to require consistent tuning no matter what stage you're at. At this scale, small fraction-of-a-percentage-point improvements are going to translate into big profits.

- At the same time, be on the lookout for disruptive innovations, to make sure that you're ahead of the pack and aware of precisely everything your competitors are doing.

Valuation

- Once again, at this scale pretty much everything goes out of the window. If you keep your eyes on growth, profits and great product, you're going to drive valuation.

Investment

- In terms of investment, things stay the same. The healthier your business is in terms of profitability and growth, the more likely you are going to be able to fund any activities from the company itself. If you still need to fund growth, then you're going to be able to raise it on attractive terms.

Chapter 36

Unicorns do Exist

Everything around you that you call life was made up by
people that were no smarter than you, and you can change
it, you can influence it, you can build your own things that
other people can use.

Steve Jobs uttered these poignant words in 1995 during an interview with the Santa Clara Valley Historical Association. Jobs's
personal journey was particularly trying. At the time, Apple was
in turmoil, and Jobs would only return to Apple a year later in
1996.

What Jobs says next in the interview is particularly powerful:

... shake off this erroneous notion that life is there and
you're just going to live in it versus make your mark upon
it. Once you learn that, you will never be the same again.

That's one of the key messages in this book. Life is about building
your vision; it's about being an entrepreneur. It's not about just
existing within the world: it's about helping to shape it.

Is that an easy process? Is it straightforward to build your vision into a billion-dollar reality? Well, the real answer is no. (Did you really expect it to be yes?) But that doesn't mean that achieving something enormous is not possible – and it sure as hell does not mean that you shouldn't try. It's worth remembering that the world was created by people just like you.

Take the story of Brian Acton. In May 2009, he applied for a job at Twitter. He was turned down.[1] Later that year he applied for a job at Facebook. He was rejected.[2] Things didn't look too bright. Later that year he partnered with a friend, Jan Koum, and founded a company. Not even five years later, in February 2014, that company was acquired for a record $16 billion. Funnily enough the company that acquired them was Facebook. Their company was WhatsApp.

Great entrepreneurs are the people who never say die, who are never defeated. They are the ones who persevere – the successful ones are the ones who persevere wisely. It's about understanding what isn't working and adjusting, tuning and changing – and finding a way that does work. It's not about forcing things: it's about finding out what works.

This section is about the breakthroughs – about the final steps – that only the best companies, and now the best apps, are able to make. What differentiates the super-successful apps from the ridiculously successful? I've trawled through all the data so you don't have to. There are truly some amazing things that the best companies do – and even invent – to cement their place at the very top.

Chapter 37

People at a Billion-Dollar Scale

Top founders – the ones who have built billion-dollar apps and billion-dollar companies, and who have remained at the top of their sector – focus massively on culture. Why is that?

In the world of fast-moving technology your advantage is not the hardware (that's owned by Apple or Samsung or someone else): rather it is the software you create, the experience and emotion that the software delivers, and the supporting experiences, such as customer service.

All of this is created and delivered by people. Not just engineers, designers and product people, but also people in call centres, the support teams answering emails and the people within the marketing, operational and administrative roles. Making sure that each and every one of these people is excited to get up and come to work every day is what great CEOs of great companies do. And they accomplish that with culture.

Culture at Scale

You need to be deliberate in creating – and then preserving – your culture, even in the early days. Your culture is the way people operate and how decisions are made – like sharing information and opinions up, as well as down, the hierarchy. This needs to be continually reinforced. People need to be empowered – told they can and should contribute to the health of the company.

Fostering a positive culture in your company will create employees who live and breathe your company mission – they are the people who infect others with the vision, who stay late, who come up with ideas, who go beyond all levels that a normal person would and who will be instrumental in helping your company to achieve billion-dollar success.

During the early days of Hailo, we adopted the culture of heavily using our app – and being hard on ourselves. Everyone in the company could hail unlimited taxis on the app – but only if they gave feedback via the app. The feedback needed to consist of both a statement of how the app could be improved and an interview with the taxi driver using the service. This led to mountains of great feedback, and a feeling that everyone in the company was actively involved in developing the product. This philosophy further extended to all the candidates we recruited, who needed to have used the app before their first interview. If they didn't show the initiative to use it, comment on it and desire to improve it, we'd cancel the interview. This bred a very powerful loyalty and pride within the company, whereby everybody constantly looked to improve the experience and actively solicit feedback across all the cities we operate in. It led to more than 10,000 five-star reviews on the app stores.

Aaron Levie, CEO of Box – an online and mobile collaboration and storage service, with a great mobile app, valued at over \$1 billion,[1] gives an example. He had a brand-new employee who noticed something wrong on a marketing programme the company was doing. They were talking about it and Levie said, 'Well, why don't you just go tell the person who's running the marketing campaign that you don't think it's a good idea?' And he responded, 'I didn't know that would be appropriate.' Behaviours need to be clarified and enforced.

For some companies the culture means staying up all night. At other companies, such as SurveyMonkey, it means the CEO goes home at 5.30 p.m. to spend time with his kids. SurveyMonkey is a billion-dollar Internet company (and the world's largest survey company[2] with over \$100 million in revenues). It developed its own culture, one focused on supporting employees with family who prefer to work more reasonable hours. This is not something you might expect in San Francisco. CEO Dave Goldberg (who also happens to be married to Sheryl Sandberg, the COO of Facebook) didn't want to take the company public, so that they could maintain their culture.

According to Goldberg, 'You need a mixture between the two [experience and culture]. Some of the worst mistakes that I've made in hiring come from hiring someone who looks like they've had great experience, but they just didn't fit with our culture.'[3]

Goldberg continues, 'There are people who don't have any experience but are just really smart, talented and motivated. When you get those people right, they're your "homegrown talent", if you will. These people are your farm team. These people are, for the most part, the best people who will stay long-term at your company. They're the carriers of the culture. They grew up there. You took a chance on them. They've learned how to be in the business.'[4]

And the newest wave of app companies are chanting the same mantra. Ilkka Paananen, CEO of Supercell, the Finnish app game maker behind Clash of Clans, says, 'If we want to create the best products, we need the best people. That was actually how the whole thing started, and where the name Supercell comes from. We're creating these small but ultra-dynamic teams of developers who work relatively independently. And, despite the small size of the group, we have big dreams – hence Supercell.'[5]

Supercell does practise what it preaches – it chases great talent very hard. On average, its employees have 10 years of experience in shipping commercial games. That affects Supercell's ability to grow as fast as its competitors – but, frankly, given its financial performance, that's not a big concern.

'We think that the biggest advantage we have in this company is culture,' offers the industry veteran. 'We want to build a very different type of company. At the centre of it is this idea of small – if you think around the console industry, or even if you look at newer platforms like Facebook, what happens is that somebody comes in, and they have this small and very passionate team, and they make a great game, and consumers pick it up.'[6]

Paananen explains, 'We don't have dedicated game designers as such – it's the team that is going to build the game, and they are all responsible for the end-user experience.'

Supercell's outrageous success is evidence that this approach can drive huge success. 'People really step up and take more responsibilities,' adds Paananen. 'It's a lot more motivating to do that, and a lot more passion gets thrown into the product.'

That is what building culture is all about.

Perks of the Job

This all does seem a little crazy, but the top Internet and app companies give their employees benefits.[7] Do these perks make you salivate? Well they should: they are designed to win and keep the best talent.

Let's have a look at what are considered the standard perks (offered by top companies such as Google, Dropbox, Uber and Evernote – all in the Billion-Dollar Club):

- Free snacks and free coffee – that's a staple for a tech startup
- Full-blown (always free) restaurant-quality cafeteria (Google adopted this in a big way)
- Free beer on tap – most people don't drink it during the day but it's nice to have at the Friday team meeting
- Company swag such as T-shirts, hats and mugs
- Discounted or free gym membership
- Dog-friendly office
- Ping-pong table, foosball table or pool table – I've seen these in many a startup office.

So that's just the standard stuff. How about some of the really cool benefits?

- Uber, the personal-chauffeur app, offers employees free, unlimited car service. They can even use Uber to commute to work.
- Evernote offers unlimited vacation days and gives employees $1,000 to cover flights. Why? CEO Phil Libin hopes the break from work will leave employees feeling refreshed for the job.
- Tumblr provides yoga classes for free.
- Dropbox gives its workers the option to build or buy their own dream computers. The office also has a complete music studio with drum set, free whiskey on Fridays, and even laser tag!

But is this a waste of money? That's not the way top tech companies think. All these benefits and perks are investment in the happiness – and therefore the quality of the output – of their teams. Making it truly fun and enjoyable to come to work can improve efficiency and morale massively.

And one of the key reasons that leading billion-dollar companies are able to invest in this is precisely the quality – profitability – of their business models. So, when you think about it, it's not really that crazy at all.

Revenue per Employee

When companies really hit scale, and their business models are firing on all cylinders and money is really flying through the door, more options appear. The more profitable companies are, the more they can invest in their people – but also in staying at the top.

Reaching the Billion-Dollar Club – and staying there – is massively influenced by your business model. And your business model drives your profitability, which controls two very important things: first, how profitable you are determines how much you can pay your employees; second, it determines how much time (and money) you can invest solving big problems. But more on the second part a bit later.

This is one of the most fascinating metrics I have found, and something worth thinking about as you scale your company. It's all about productivity, and in the technology and software world the single biggest thing to think about is your employees.

Let's dig into the numbers behind some of the most established players – and some of our billion-dollar apps. All the data below is for the year ending 31 December 2013.

NETFLIX. Interestingly, Netflix takes the top spot with $2.2 million per employee for its video-streaming and DVD service.

APPLE. No surprises here from the most valuable company in the world – it generated a whopping $2.12 million per employee in 2013 (up from $1.3 million in 2011). But it does sell some pretty expensive hardware.

GOOGLE. The search and advertising hero is close to the top with $1.292 million in revenue per employee.

FACEBOOK. These guys are generating $1.2 million per employee per year[8] ($7.8 billion in revenues with 6,340 employees).

MICROSOFT. Microsoft is still close to the top with $786,000 per employee for its operating system and desktop productivity software.

AMAZON. Despite having a massive workforce of 117,000 the company still generates $634,000 per employee in revenue for its online super shopping mall (though its profits are not that healthy).

SUPERCELL. This 150-person company made $894 million in revenues in 2013.[9] This puts the number at $6 million per employee. Mind-boggling.

UBER. In 2013 Uber processed around $900 million in black-car and taxi bookings, which is really about $120 million in revenue (it keeps around 15 per cent of the fare amount). Its worldwide team is around 350 people, so that translates to about $342,000 per employee.

So, if you want to play in the big-boy club of technology companies, the average revenue per employee is $593,000.

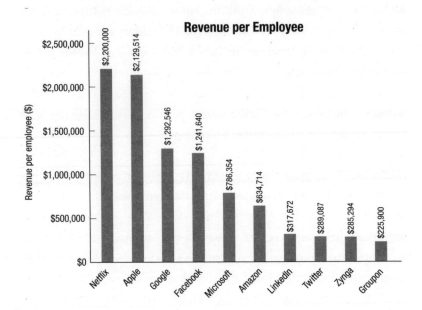

Taking this investigation one step further, how do these Internet companies compare with the broader world of technology companies?

Cisco is head and shoulders above the others at around $647,000 per employee, IBM generates $229,000 per employee (it has 434,000 employees) and HP pulls in $352,000 per employee, again due to its behemoth size of 317,000 people. It is clear to see that the advantage is maintained by having a smaller, more skilled-based team generating revenue.

Profit per employee

Let's take this analysis one level deeper. Profit is the real driver of any business – after all profit is the cash left in the bank at the end of the day that you can actually play around with.

From the chart opposite, you can see that Apple once again comes out on top with $460,000 in profit generated by each employee, and Google and Facebook are neck and neck with $280,000 and $236,000 respectively. If we added Supercell to this picture we'd get another unbelievable number. No matter how much they pay every one of their 150 employees, they are still going to be left with *millions* in profit per employee.

So the mechanics of mobile apps – combined with the distribution power of Apple's App Store and Google Play, and the simplicity of in-app payments – have created a revenue-and-profit juggernaut like nothing else.

This means that these leading companies have a lot of cash they can spend to make their employees happy – everything from the super perks we just read about, to great offices, to company holidays and events. More importantly, these companies have the means to pay big salaries to attract the very best talent. And, since they are already leading the pack, this tends to create a virtuous circle.

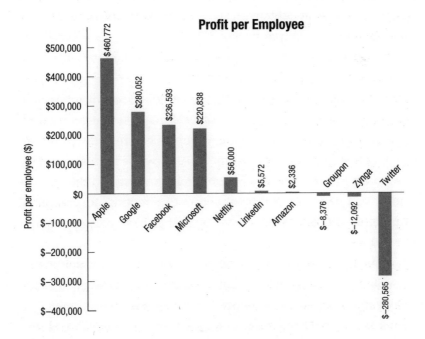

If these top companies continue to invest wisely, they have the opportunity to create a very difficult-to-beat advantage. So you can see why focusing on a great business model from the very onset is critical to every aspect of creating a sustainable business.

Moonshots

When Larry Page – Google's current CEO – was an undergrad at the University of Michigan, he joined a summer leadership course called LeaderShape, which promoted the tagline 'a healthy disregard for the impossible'. That same phrase now embodies what he sees as Google's mission.

This is an incredibly clever – and necessary – strategy for Google. In a company where people are the most important asset, and where retention is key, ensuring that people see meaning in what they're doing is the most important thing the CEO can deliver. Google already invests heavily in top salaries, great offices and facilities and super perks. But superficial investment is not enough to stay truly great. Creating a core of innovation, where people can innovate, create and disrupt, is what is required.

Larry Page wants to encourage his employees to be innovative to the very core. And to communicate that to engineers – well, basically, you need to put some numbers behind it. Anyone can deliver a 10 per cent improvement. And, while that is necessary and useful for the business, it isn't going to result in the innovations you need to stay number one. Shooting for marginal improvements is also not the vision that gets people jumping out of bed in the morning.

What is interesting, on the other hand, is improving something tenfold.[10] Achieving that kind of result is true cerebral gymnastics. Delivering 1,000 per cent improvement requires rethinking problems from first principles; it requires that people go out and explore the frontiers of research and development; it means a lot of challenges, and fun in the process.

When you think about it, this is nirvana for knowledge workers. Engineers, developers and product managers are all problem solvers. By telling them that their company is a playground – a really well-kitted-out playground designed for them to go and solve gargantuan, meaningful problems – well, that is the ultimate retention tool.

And my point from earlier is that Google's revenue – and profit – per employee allows it to pursue this strategy. Companies that don't have robust business models will not be able to invest in these kinds of activities, which will make it increasingly harder for them to retain the best people, who in turn, once salary is taken care of, will be looking for a job with meaning. And that comes from a company that has a culture of pure innovation and solving meaningful problems.

Google X is the division of Google that is home to the company's moonshots. Since 2010 it has delivered a variety of seemingly impossible fantasies, such as the self-driving car (which has travelled over 500,000 km without a single accident[11]), Google Glass (a wearable computer with an optical head-mounted display), Project Loon (which provides rural Internet connectivity via high-altitude autonomous balloons[12]) and more than 100 other projects.[13]

So, when you think about the future of your app, it's important to think about how big your ambition and vision are – and how you are going to take people on that journey. It's not just a journey about earning lots of money and getting lots of perks. At the end of the day, to create a sustainable, long-term company you're going to need to have a mission infused with meaning that keeps your team inspired and excited to come to work every day.

Chapter 38

Advice from Billion-Dollar CEOs

'It's about making *mostly* the right decision
all the time *more quickly*'
#BILLIONDOLLARAPP

A great CEO is critical to delivering a billion-dollar vision. Let's have a look at some of the biggest lessons learned. All the examples here come from people who have done it – they have achieved billion-dollar greatness.

I've collected the advice from the CEO of the top mobile-app companies and then also Facebook, Amazon, LinkedIn and Google. This is the kind of advice you should listen to and is relevant to every stage of your journey of building a great app.

Learning to Let Go

Bill Gross knows a thing or two about startups – he's had 35 IPOs and exits and 40 failures. He was also the guy who came up with paid search – the business model that Google 'adopted' to transform itself from purely a great piece of technology into a great company.[1]

Gross seems OK, even happy, with how things turned out. He raves about Google. 'I'm wildly proud of coming up with the paid-search model,' he says. 'I didn't know how big it was at the time.'

Besides, Gross says, if Google didn't make billions with the pay-per-click auction model, it would have made its billions some other way. 'I wish I had come up with the Google idea,' he says. 'The Google idea was the idea for organising the world's information. Mine was just an idea for making money.'[2]

One of the big messages from Gross is about spreading load at a high growth. You cannot do everything yourself – and, frankly, you don't want to do everything yourself. It's about delegation, not becoming a bottleneck because it's *your* company. That causes massive friction and stops your company from scaling.

Another lesson in delegation comes from Drew Houston, an MIT and Y Combinator graduate and CEO of a billion-dollar company, Dropbox, who has a brilliant mobile app with more than 100 million users. 'The hardest transition was going from working directly on the product to not,' he says.[3]

He goes on to say, 'There's this joy that comes from sitting down to solve a problem and standing up when it's done and good. Building a company or managing people is never just done.'

Make Decisions – Fast

Bill Gross also talks about speed. He suggests, 'It's about making *mostly* the right decisions all of the time *more quickly*.' Don't get caught up on making perfect decisions, but keep the momentum going. It's better to make a decision quickly that's 80 per cent right than a decision 100 per cent right too slowly. Especially in the mobile age, things change too fast and your decision is very time-sensitive. What is right today may no longer be right in six months. Be nimble.

Larry Page, Google's CEO, echoed Gross at one of his company's annual Zeitgeist conferences:[4]

'One of the interesting things that we've noticed,' he said, 'is that companies correlate on decision making and speed of decision making. There are basically no companies that have good slow decisions. There are only companies that have good fast decisions. I think that's also a natural thing as companies get bigger – they tend to slow down decision making. And that's pretty tragic.'

Vision and Mission Statements

Having a clearly articulated 'vision' and 'mission' is critical to guiding your company to an ultimate goal. They set the foundation for the company's strategy, and then its objectives. Most people tend to use the words 'vision' and 'mission' interchangeably, points out Jeff Weiner, the CEO of LinkedIn (which is currently worth about $20 billion but does only about 5 per cent of traffic via mobile[5]).

'Vision is the dream,' Weiner points out. 'A company's true north. It's what inspires everyone day in and day out.' The LinkedIn vision is 'Creating economic opportunity for every professional'. Everything that everyone does – whether it be designing, building, programming or marketing – should help deliver on the vision.

The 'mission', on the other hand, is *how* the vision is executed. At LinkedIn that means 'Connecting the world's professionals to make them more productive and successful'.

Visions are meant to be grandiose – multi-decade adventures that may seem close to impossible to deliver. That's why they are called visions. The more meaning the vision has, the more soul is injected into the company.

Google has a brilliant mission statement: 'Google's mission is to organize the world's information and make it universally accessible and useful.'

Weiner admits that most startups have only a vision or a

mission statement, but he points out the benefits of having both the 'what' (vision) and the 'how' (mission) working together. The CEO needs to ensure these are in place, to ensure there is a real purpose for the company, which is a mechanism to both inspire and organise employees.

Leading gaming company Rovio has a mobile-centric mission statement: 'We provide unique and novel stories with innovative game-play to satisfy the growing demand of games designed for mobile.'[6]

Vision and mission statements evolve over the lifetime of a company, so don't worry if you don't settle on one immediately. They should come organically; they need to feel right. Google didn't flesh out its mission statement until the company was about three years old. But when that mission was crystallised it really injected a meaning and purpose into the company, and allowed it to expand into so many new and fantastic areas, all which really do seem to make sense.

Invest in Diverse Skills

Bill Gross has backed and advised 100-plus companies and believes that you fundamentally need to build your team with four types of complementary people[7]:

- **Entrepreneurs** – the ideas people who explore what it takes to get ideas off the ground;
- **Producers** – they are the doers who push projects forward and deliver; they make the product, sell it and answer customers' questions;
- **Administrators** – these people are builders, who plan, organise and devise processes; they keep your organisation running smoothly as you grow; and

- **Integrators** – they are the emotional centres that make sure that everyone gets along, especially in times of high stress and pressure.

In an ideal world you want these four viewpoints when your company is making big decisions, and ideally they will be represented on your senior management team.

Have Staying Power

Today, Pandora is a billion-dollar music-streaming company, and one of the most used apps in the world. But, back in 2000, at the end of its first year, the company was running out of money.[8] Founder Tim Westergren failed to raise any additional funding and was forced to shrink the team to 50 and didn't take a regular salary for the next two years. Pandora endured for a number of reasons: users loved the product; and Westergren was totally transparent with his people no matter how bad things got, carefully handpicking his initial team for staying power rather than attracting people who would flee in a crisis. During that time he led by example: he was the first one in and the last one to leave. He also paid salaries with his own credit cards. The journey to success for Pandora was a long one, but that's how success happens: the perseverance of the cofounders.

Learning About Leadership

When Dropbox started to take off, Drew Houston, the CEO, started reading. He knew that he had a lot to learn and relied on any resources he could lay his hands on. He knew he had to adapt quickly if he was going to make his company a success. Y Combinator – the accelerator programme – helped the young CEO along the way.

'It's a mix of a variety of different things,' he says, 'like mentors – people who are experienced and have been through the process many times – and peers. Some of my best friends are Y Combinator founders, and we all went through the same kind of thing at the same time.'

Dropbox's early board of directors also helped him recognise patterns and gave him an idea of what he should be thinking about. In the end, his attitude proved to be the most important resource to draw on.[9]

'The advice I would give is to get used to things,' says Houston. 'Get out of your comfort zone with things even as straightforward as public speaking. It's a mindset – spending a lot of time thinking about what you don't know that you don't know. Those are questions to ask: "What am I going to need? What is the company going to need? What should we be doing 3, 6, 12, 24 months from now?" If you can back that up to today, that can be a useful lens to figure out the skills you need to develop.'[10]

No Desperate Hires during Rapid Growth

Aaron Levie is a young founder who managed to scale Box into a billion-dollar business without compromising its culture.[11] That's immensely hard. He attributes much of this success to keeping the bar high on hiring, no matter how pressing the need to fill a role. To limit 'desperation hires' he instituted what he called the 10-person test. 'For anyone you hire, would you want them to be in the first 10 people at your company?' he asks. 'If you're hiring the 400th employee, you would still want that person to be in your original 10.' It's impossibly hard to stick to this rule in my experience. Outside Silicon Valley – which has a unique density of talent that's not available anywhere else in the world – it's hard to keep such impossible standards. However, the better the people you hire, the more productive they are, which optimises how many people you need to hire. Google has a very similar philosophy, which has allowed it to be one of the most talent-dense companies in the world.

Be Prepared, Not Lucky

'When you get lucky, it's because you didn't know what was going to happen. The corollary is, if you know what's going to happen then there is no luck. There is no uncertainty and there is no risk.'[12] The words are those of David Friedberg, a former member of the corporate development team at Google and founder and CEO of the Climate Corporation, who sold his company in 2013 for $1 billion. All businesses naturally come with a degree of risk. You don't know where your market is headed, what your competitors are working on or even if people are going to use and pay for your product. So take a more methodical approach: identify all the uncertainties, research the hell out of them, and then deal with them systematically so they are no longer uncertain. 'Identify the unknown, mitigate the unknown,' Friedberg says. 'Only then can you enable the outcomes you want and increase the value of your company.'

Open Communication

Key to the success of any company is transparency and trust. Jack Dorsey, Square's CEO, is insistent that everyone at Square should know what the company is up to and why it's doing it. So he instituted an astonishing rule: at every meeting involving more than two people, someone must take notes and send them to the entire staff.

It doesn't matter what the meeting is about: bug fixes, new partnerships, pending contracts, a new launch, important metrics. Everyone hears about them. Dorsey says he often gets 30 to 40 meeting notes every day. He filters them in his inbox and reads them through on his iPhone when he gets home at night.

Even the structure of your office is important: encouraging communication and circulation is good for creativity. 'We encourage people to stay out in the open because we believe in serendipity – and people walking by each other teaching new things,' says Dorsey.[13]

Jonathan Rosenberg led Google's product strategy for almost 10 years. He has a lot of rules for great managers, but the most important element of leadership needs to be transparency. 'At Google, our default mode was to share all information,' he says. 'We strived to empower everyone equally. In the Internet age, power comes from sharing information, not hoarding it.'

Employees like being trusted and hate being surprised. A policy of complete transparency feeds these needs. Rosenberg's advice is, 'Back up your position with data. You don't win arguments by saying, "I think." You win by saying, "Let me show you."'[14]

Chapter 39

Getting Acquired

'Twitter's IPO created 1,600 new millionaires in one day'
#BILLIONDOLLARAPP

A wise person once told me that companies are never sold: they are acquired. There's great truth in that statement. All the great entrepreneurs in this book were not looking to furiously grow and 'just' sell their companies. They had much bigger visions, and ultimately an appropriate partner became entangled in that vision and wanted to help realise it.

Each year in the US between about 400 and 700 venture-backed companies go through an exit event. The vast majority of those – about 75 per cent to 90 per cent – are acquisitions.[1]

Google has made an art of acquiring companies – big and small. This talent has allowed Google to grow in leaps and bounds, allowing it to stay ahead of the competition, but also building a formidable talent pool. It has 80 specialists across the organisation to make the process as smooth as possible.[2]

Google's acquisition of YouTube was the foundation of its video strategy; its acquisition of Keyhole Inc. was the foundation of Google Maps; its acquisition of DoubleClick launched it into display advertising; its AdMob acquisition further cemented its lead in mobile advertising; and even Android – the world's biggest mobile operating system – was an acquisition.

So let's hear some advice from Google 'acquisition chief', David Lawee, VP of corporate development:

> The most important thing I look for is alignment between what the entrepreneur wants to do with their product and their company and what Google wants to do. If there is perfect alignment, then it has a very high chance of success. If there is not, then we should not be doing it.[3]

It's also important to have a big vision:

> Go big. When your vision is large in scope, it's easier to get the capital you need, the top-shelf employees you want, and ultimately, the acquisition offer you hope for.[4]

Waze

In June 2013 Google completed its first billion-dollar-app acquisition. It was an Israeli social-mapping and traffic app, Waze (Facebook and Apple were also rumoured to be in negotiations to acquire the app[5]).

Waze built up a fast and loyal following, amassing 50 million users in 190 countries.[6] The company built its own mapping technology from the ground up, so was not at all reliant on Google, Apple or Microsoft's own mapping technologies. The app was also engineered to be social from the ground up, users being able to chat to other Waze users, and focused on user-generated traffic data, as well as petrol prices.

The team was 80 strong – 70 based in Israel, working on the product and engineering, and a small team of 10 in Palo Alto. The acquisition actually totalled $1.15 billion – $1.03 billion went to the company's investors and $120 million was earmarked as retention payments for employees, who on average received $1.2 million each before taxes.[7]

Within a few weeks of the acquisition Google was featuring Waze crowd-sourced traffic information in Google Maps – a testament to how efficient Google's integration process really is.

The Best Defence is Offence

The body-numbing price tag of $16 billion[8] (plus another $3 billion in incentives) forked out by Facebook to acquire WhatsApp in early 2014 seems beyond the realms of lunacy on the surface. What's more interesting is that it happened at true Internet speed, with Zuckerberg revealing that it took a mere 10 days from suggesting a possible acquisition deal with WhatsApp's CEO Jan Koum to publicly announcing the deal on February 19.[9]

That perceived lunacy is rooted in some rather interesting numbers:

- WhatsApp has about 300 million active daily users,[10] Facebook has 728 million (as of February 2014).[11]
- WhatsApp's messaging volume is approaching the entire global telecom SMS volume.[12]
- In the words of Mark Zuckerberg: 'WhatsApp is on a path to connect 1 billion people. The services that reach that milestone are all incredibly valuable.'[13]

But the real kicker was WhatsApp's growth rate. Nothing in history has grown faster. Given the app's momentum it could become a massive threat for Facebook, which is already seeing other apps (like Snapchat) stealing users' attention. So at an acquisition price that represented almost 10 per cent of Facebook's market capitalisation, perhaps it was a small price to pay.

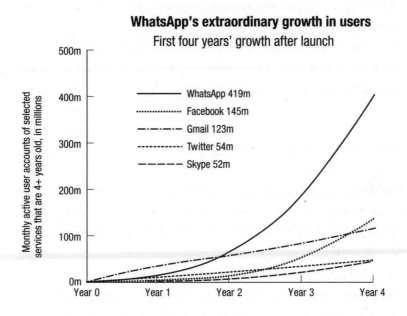

WhatsApp's extraordinary growth in users
First four years' growth after launch

Buying Sustained Growth

One of the biggest reasons that companies acquire is to drive growth. Sustaining growth is critical to any company's success – especially that of a technology company. And, in an industry where it's very hard to predict what will be successful, you need to be eternally vigilant.

Facebook acquired Instagram for a cool $1 billion to get its hands on one of the fastest-growing social photography apps. It was clear that Instagram's then 30 million users were uploading so many photos that they were encroaching on the volume being uploaded by Facebook's own users.

People's attention is a limited commodity. If a new app is stealing people's attention in a big way, that means it is stealing attention away from another app or website. And that means that someone's precious advertising dollars are going to decrease.

I've talked about the explosive growth of Snapchat before – Facebook tried to acquire the app for $1 billion,[14] but was flatly turned down. Facebook tried again in November 2013, this time offering a reported $3 billion.[15] It was flatly rejected again. Facebook's business is monetising our attention via advertising – and increasingly that's mobile advertising. With Snapchat usage of 150 million[16] snaps (shared photos or messages) per day eclipsing that of Instagram at 40 million photos daily, and approaching the 300 million photos shared on Facebook daily, Facebook could be in big trouble. It's only a matter of time before Snapchat traffic eclipses that of Facebook. And that's going to be a huge issue.

Yahoo! is following precisely the same strategy as Facebook, and in May 2013 the company forked out $1.1 billion[17] for Tumblr, the super-simple blogging platform used by some 170 million users.[18] Tumblr already has a large mobile audience – with mobile growing three times as fast as desktop traffic.[19] Tumblr also tends to have a much younger user base – something that is increasingly attractive to advertisers.

So there will always be a great – albeit risky – opportunity to create an app that captures and holds people's attention. If you can build something truly sticky, there will always be someone waiting in shadows with the ability to monetise it.

Non-Tech Corporations are Eating Startups

Google, Facebook, Twitter and Yahoo! are seeing the competition heat up in the race to acquire the best software startups.

Marc Andreessen is the cofounder and general partner of venture-capital firm Andreessen Horowitz, which has invested in Facebook, Groupon, Skype, Twitter, Zynga and Foursquare, among others. He is not alone in saying, 'We believe that many of the prominent new Internet companies are building real, high-growth, high-margin, highly defensible businesses.'[20]

As Internet startups build more robust businesses, faster, they are becoming highly prized acquisition targets.[21] Non-tech companies are now making billion-dollar acquisitions that no one would have dreamed possible. Monsanto acquired Climate Corporation[22] for a whopping $1 billion in late 2013. Monsanto focuses on providing seeds, biotechnology and crop-production products for farmers around the world. Monsanto plans to use the acquisition to leverage its big-data expertise to optimise farming globally.

The insurer UnitedHealth Group bought the health-data analytics company Humedica for a few hundred million dollars. The fitness-clothing retailer Under Armour bought the fitness tracking-app developer MapMyFitness for $150 million. Office-supply retailer Staples bought e-commerce personalisation company Runa. First Data – an old-school player in the payments industry – acquired both mobile-loyalty startup Perka and mobile-payments startup Clover within a year. Even Target, the retail behemoth, has splashed out to acquire a number of e-commerce companies. Ford Motors picked up an in-car music app startup called Livio.[23] And there are plenty more examples.

There's even a newer startup that helps broker introductions between startups and acquirers, called Exitround. It has seen strong interest in non-tech corporations joining the marketplace and actively seeking acquisitions.

'Their main motivation is realising that software is eating the world, and they have to add software talent and technologies to their products,' explains Exitround founder Jacob Mullins.[24]

Mullins says that 20 per cent of acquirers on his site are public companies and 10 per cent of those are Fortune 500 companies. All this is great news for entrepreneurs – it can only create a more dynamic and exciting set of exit possibilities.

It's worth keeping in mind, however, that the majority of technologies fail and don't reap the benefits that the acquiring companies hope for. The complexity of integrating entrepreneurial culture into a new corporate culture, aligning different technology platforms and products, is troublesome for even the most agile tech companies. So be aware of the reality of what might happen post acquisition and prepare yourself for it.

The IPO

At the time of writing, only one app (Candy Crush Saga from King) has ever gone through an IPO – or initial public offering. An IPO is when shares in your company are sold to the public on a securities exchange such as the AIM or the NYSE for the first time. Through this process your company, which up until now has been private, transforms into a public company.

IPOs are used in the case of technology companies to monetise the investments of early investors; in this case the investors can sell their shares directly to the public. Companies can also use an IPO to raise further capital for expansion – this time getting money from investors on a stock market, rather than through venture capitalists.

In any given year about 7 per cent of exits are done via the IPO route – though this does vary wildly. In 2013, for example, it represented 15 per cent.[25]

The IPO process is fraught with cost and complexity, so is not really a viable option for smaller companies. It also comes with competitive risks as well. Because of the reporting requirements and disclosure of certain types of information, it can help out your competitors. At a certain scale, however, it becomes the best option.

Pitfalls of technology IPOs

Only the most robust startups have made it to this level. Facebook is a great example. It held its IPO on 18 May 2012. The IPO was one of the biggest in technology, and the biggest in Internet history, with a peak market capitalisation of over $104 billion and a share price of $38.

Within days, however, the share price plummeted and, two months after the IPO, it hit $20 and lost half its valuation. The public stock market wasn't ready for Facebook, but the company worked hard and 16 months later the share price had rebounded to the IPO level, based on strong growth of users, revenues and profits. At the end of 2013 it was trading at 25 per cent above its IPO price. It was a wild ride.

Twitter goes public

Twitter – the micro-blogging service – announced in September 2013 that it was going to an IPO. It seems as if the response has been unanimously positive, which is an encouraging sign for mobile-first social media. On the first day of trading, Twitter's shares closed at $44.90, giving it a market capitalisation of nearly $25 billion. The IPO process allowed the company to raise about $2.1 billion, which it can now use to fund further activities.

The day also saw a huge windfall for employees: the IPO created 1,600 new millionaires.[26]

Unlike Facebook, which held its IPO when the lion's share of its revenue still came from ads on desktop PCs, Twitter has begun its public life as a company that draws 65 per cent of its revenue from mobile ads. So this is really laying the groundwork for a new wave of mobile-centric, and even mobile-first, companies with robust business models to also start looking at the IPO route.

The First Mobile IPO

King is the British company behind the hits Candy Crush Saga, Bubble Witch and Pet Rescue Saga and it is the first app gaming company to file to go public on the New York Stock Exchange.[27] It held its IPO on 26 March 2014 with a valuation of $7 billion. At the end of 2013 Candy Crush had been downloaded more than half a billion times across both desktop and mobile, and been played more than 150 billion times. King's 2013 sales topped $1.88 billion, up from about $164 million in 2012.[29]

But King was founded back in 2000, and has only recently become a mobile giant. It isn't really one of the new wave of mobile-first companies. The top contender for the first app-first company to IPO has to be Uber, everyone's private driver. Having raised a whopping $258 million in funding in 2013, it is essentially cash-flow-positive, so it doesn't have any pressing need to go public. And, since it was founded as recently as 2009, there isn't any investor pressure to get their money back yet.

It seems like Uber has more operational goals on its mind for the moment. With more than 1 million riders, the company hopes to roll out its service in the 500 biggest, global cities, according to its CEO Travis Kalanick.[30]

Wrapping Up

There has been a real step change, not only in Silicon Valley but also with entrepreneurs around the world. In 2000, when I graduated from college, the Internet bubble was just beginning to pop and scepticism about technology startups was abundant. It took the better part of a decade for the world to figure out how to build valuable – and durable – Internet companies.

The best entrepreneurs have cracked the formula, and now,

with more experience under their belts and the massive innovations in technology, platforms, business models and distribution, we are already seeing a new wave of mobile-centric creation. And it's just the beginning.

That formula has a number of elements. The first is to build a truly great product – an app, a service – that addresses in an elegant way a big problem that many people share. The raw computing power, coupled with Internet connectivity, supported by the API economy, is empowering even rookie developers to create spectacularly powerful apps. The key is spotting an opportunity, and focusing on building something great for users. Don't ever listen to people who say it's good enough, because it never will be. Building a great product is a fluid, ongoing process.

Today, the best entrepreneurs are focusing on products and apps that are materially affecting billions of lives. Their apps are making lives better, more enjoyable, more social and more efficient. Technology is being tamed; it is becoming more about people. Technology in many senses is disappearing into the background – it is increasingly woven into our lives, our experiences. Software – apps – is personal. Building apps is fundamentally about understanding people.

The second secret to success *is* people. In the virtual realm of software and apps, the quality of the end experience you offer your users is a direct reflection of the quality of your engineers, designers and product managers. Great, tough people are more suited to the ups and downs of startup life than average people. If you can infect people with your vision and mission you will succeed. Remember that it's not your job to make everyone happy: it's your job to attract and select the best people for the mission. You should be interviewing 20 to 40 people before you make a hire. You should be demanding and selective. You're building a company, a culture, a group of people with a relentless mission to solve a big problem. Make sure you absolutely want every person on your team with you.

Possessing a solid business model from Day One is the third key to success. Whether you're focused on building a gaming company with in-app payments, an e-commerce marketplace, Software as a Service, enterprise subscriptions or – the toughest – building a consumer audience app that will be later monetised via advertising, you need to have a concrete idea in place. It's just not good enough to fumble around looking for a business model. Yes, some people trip into it, but the odds are very low. The arrival of app stores, in-app payments and super-simple credit-card payment integration (from players such as Stripe) has made it so easy to collect money that you should be doing it from the day your app goes live.

The last magical factor is you. Malcolm Gladwell's book *Outliers* suggests that excellence is the product of *practise* more than innate genius. Our billion-dollar app entrepreneurs are definitely human – they've made plenty of mistakes but they persevere, they keep going. But it is their unique way of focusing on one thing they are so passionate about – that they so deeply believe in – that makes the difference. These are not people driven by megalomaniacal power. They are people driven to make the world a better place in the way they can. Similarly, if you find a cause you're passionate about, you have a chance to push through all the barriers, all the naysayers and all the inevitable challenges to make your app a success.

It has been eye opening getting into the heads of the billion-dollar-app founders. They are impressive people characterised by their energy and vision. When they talk about the future you can see the possibilities in their eyes, their enthusiasm about what is coming next and an infectious sense of urgency to get there now. They seem painfully different from the traditional corporate CEO in his penguin suit, placating shareholders and painfully removed from the rank-and-file employees.

Mobile is just the beginning of a new wave of technology – and therefore opportunity. More than a billion people around the world already actively use smartphones – and apps – daily. Within just a few years that will double and triple.

We are alive in a time when technology has never been as powerful – but it's also never been as accessible. It's up to you to find that idea and hone it, and then find the passion that will take you on your own journey to building something truly great.

> Your work is going to fill a large part of your life, and the only way to be truly satisfied is to do what you believe is great work. And the only way to do great work is to love what you do. If you haven't found it yet, keep looking. Don't settle.

> As with all matters of the heart, you'll know when you find it. And, like any great relationship, it just gets better and better as the years roll on. So keep looking until you find it. Don't settle.

> *– Steve Jobs, cofounder of Apple*

Notes

Part I: Think Big

Chapter 1: The View from the Inside

1 A billion, for the purposes of this book, is the short billion, i.e. a thousand million.

2 'List of the Wealthiest Historical Figures', entry on Wikipedia.org, en.wikipedia.org/wiki/List_of_wealthiest_historical_figures.

3 According to estimates from the US Census Bureau. World Population: Historical Estimates of World Population, www.census.gov/population/international/data/worldpop/table_history.php.

4 'World Population Milestones', entry on Wikipedia.org, en.wikipedia.org/wiki/World_population_milestones.

5 'Facebook Reports Fourth Quarter and Full Year 2013 Results', press release on FB.com, 29 January 2014, investor.fb.com/releasedetail.cfm?ReleaseID=821954.

6 Wolfgang Gruener, 'Facebook Estimated to Be Running 180,900 Servers', article on TomsHardware.com, 17 August 2012, www.tomshardware.com/news/facebook-servers-power-wattage-network,16961.html.

7 Rich Miller, 'Facebook's $1 Billion Data Center Network', article on DataCenterKnowledge.com, 2 February 2012, www.datacenterknowledge.com/archives/2012/02/02/facebooks-1-billion-data-center-network/.

8 Rich Miller, 'Google Data Center Spending Continues to Soar', article on DataCenterKnowledge.com, 18 October 2013, www.datacenterknowledge.com/archives/2013/10/18/google-data-center-spending-continues-to-soar/.

9 Rich Miller, 'Google Has Spent $21 Billion on Data Centers', article on DataCenterKnowledge.com, 17 September 2013, www.datacenter-knowledge.com/archives/2013/09/17/google-has-spent-21-billion-on-data-centers/.

Chapter 2: Mobile Genetics

1 'ITU Releases Latest Global Technology Development Figures', press release on ITU.int, 27 February 2013, www.itu.int/net/pressoffice/press_releases/2013/05.aspx#.UwypjeN_tAo.

2 Flurry Analytics, 2013 and Mary Meeker and Liang Wu, KPCB Annual Internet Trends Report 2013, published May 2013, slide 40, www.kpcb.com/insights/2013-internet-trends.

3 Jessica E. Lessin and Spencer E. Ante, 'Apps Rocket Toward $25 Billion in Sales', article on WSJ.com, 4 March 2013, online.wsj.com/article/SB10001424127887323293704578334401534217878.html.

4 Mike Croucher, 'Supercomputers vs. Mobile Phones', 2 June 2010, www.walkingrandomly.com/?p=2684.

5 Mary Meeker and Liang Wu, May 2013, slide 109, op. cit.

6 Mary Meeker and Liang Wu, May 2013, slide 33, op. cit.

7 Data retrieved from www.cnnic.net.cn/hlwfzyj/hlwxzbg/hlwtjbg/201401/P020140116395418429515.pdf

8 Ibid.

9 StatCounter Global Stats, 'Comparison from November 2013 to January 2014', gs.statcounter.com/#desktop+mobile-comparison-ww-monthly-201311-201401-bar.

10 'Smartphone Users Worldwide Will Total 1.75 Billion in 2014', article on eMarketer.com, 16 January 2014, www.emarketer.com/Article/Smartphone-Users-Worldwide-Will-Total-175-Billion-2014/1010536.

11 Tomi Ahonen, 'The Annual Mobile Industry Numbers and Stats Blog – Yep, this year we will hit the mobile moment', blog post on CommunitiesDominate.blogs.com, 6 March 2013, communities-dominate.blogs.com/brands/2013/03/the-annual-mobile-industry-numbers-and-stats-blog-yep-this-year-we-will-hit-the-mobile-moment.html.

12 The average American aged 15 and over sleeps for 8.7 hours, as reported in 'Charts from the American Time Use Survey', 23 October 2013, www.bls.gov/tus/charts/.

13 Ingrid Lunden, 'Gartner: 102B App Store Downloads Globally in 2013, $26B In Sales, 17% From In-App Purchases', article on TechCrunch.com, 19 September 2013, TechCrunch.com/2013/09/19/gartner-102b-app-store-downloads-globally-in-2013-26b-in-sales-17-from-in-app-purchases/.

14 'Gartner Says Mobile App Stores Will See Annual Downloads Reach 102 Billion in 2013', article on Gartner.com, 19 September 2013, www.gartner.com/newsroom/id/2592315.

15 Kevin Bostic, 'Apple Refunds $6,131 iTunes Bill for 8-year-old's Unauthorized In-App Purchases', article on AppleInsider.com, 22 July 2013,

appleinsider.com/articles/13/07/22/apple-refunds-6131-itunes-bill-for-8-year-olds-unauthorized-in-app-purchases.

16 Sven Grundberg and Juhana Rossi, 'Earnings Soar at Finnish Game Maker Supercell', article on WSJ.com, 12 February 2014, online.wsj.com/news/articles/SB10001424052702304703804579378272705325260?mg=reno64-wsj.

17 'Small Retailers Reap the Benefits of Selling on Amazon, but Challenges Remain', article on InternetRetailer.com, 14 February 2014, www.internetretailer.com/mobile500/.

18 Bill Siwicki, 'It's Official: Mobile Devices Surpass PCs in Online Retail', article on InternetRetailer.com, 1 October 2013, www.internetretailer.com/2013/10/01/its-official-mobile-devices-surpass-pcs-online-retail.

19 'Mobile Commerce Comes of Age', article on InternetRetailer.com, 24 September 2013, www.internetretailer.com/2013/09/24/mobile-commerce-comes-age.

20 'Gartner Says Worldwide PC Shipments Declined 6.9 Percent in Fourth Quarter of 2013', article on Gartner.com, 9 January 2014, www.gartner.com/newsroom/id/2647517.

21 Mary Meeker and Liang Wu, May 2013, slide 31, op. cit.

21 In-Soo Nam, 'A Rising Addiction Among Youths: Smartphones', article on WSJ.com, 23 July 2013, online.wsj.com/article/SB10001424127887324263404578615162292157222.html.

23 'Americans Can't Put Down Their SmartPhones, Even During Sex', article on Jumio.com, 11 July 2013, www.jumio.com/2013/07/americans-cant-put-down-their-smartphones-even-during-sex/.

24 Drew Olanoff, 'Mark Zuckerberg: Our Biggest Mistake was Betting Too Much on HTML5', article on TechCrunch.com, 11 September 2012, TechCrunch.com/2012/09/11/mark-zuckerberg-our-biggest-mistake-with-mobile-was-betting-too-much-on-html5/.

25 Walter Isaacson, *Steve Jobs*, Little, Brown: London, 2012, p. 501.

26 'Standish Newsroom – Open Source', press release, Boston, April 2008, blog.standishgroup.com/pmresearch.

27 Richard Rothwell, 'Creating Wealth with Free Software', article on FreeSoftwareMagazine.com, 8 May 2012, www.freesoftwaremagazine.com/articles/creating_wealth_free_software.

28 'Samsung Elec Says Gear Smartwatch Sales Hit 800,000 in 2 Months', article on Reuters.com, 19 November 2013, www.reuters.com/article/2013/11/19/samsung-gear-idUSL4N0J41VR20131119.

29 Jay Yarow, 'Meet the Team of Experts Apple Assembled To Create The iWatch, Its Next Industry-Defining Product', article on BusinessInsider.com, 18 July 2013, www.BusinessInsider.com/iwatch-sensors-2013-7.

30 Macelo Ballve, 'Wearable Gadgets Are Still Not Getting The Attention They Deserve – Here's Why They Will Create A Massive New Market',

article on BusinessInsider.com, 29 August 2013, www.BusinessInsider.com/ wearable-devices-create-a-new-market-2013-8.

Chapter 3: A Billion-Dollar Idea

1 Eric Jackson, 'Why Silicon Valley Tech Wunderkids Will Only Ever Have 1 Good Business Idea During Their Entire Lives', article on Forbes.com, 18 June 2012, www.forbes.com/sites/ericjackson/2012/06/18/why-silicon-valley-tech-wunderkids-overestimate-their-own-smarts-and-abil-ities/.

2 Kevin Rose, 'Foundation: Evan Williams on Hatching Big Ideas', article on TechCrunch.com, 28 June 2013, TechCrunch.com/2013/06/28/ foundation-evan-williams-on-hatching-big-ideas/.

3 Evelyn M. Rusli and Douglas MacMillan, 'Snapchat Spurned $3 Billion Acquisitions Offer from Facebook', blog post on WSJ.com, 13 November 2013, blogs.wsj.com/digits/2013/11/13/snapchat-spurned-3-billion-acquisition-offer-from-facebook/.

4 'Human Universals', entry on Wikipedia.org, en.wikipedia.org/wiki/ Human_Universals.

5 Find out more about the app at www.bible.com.

6 Alyson Shontell, 'With 100 Million Downloads, YouVersion Bible Is A Massive App That No VC Can Touch', article on BusinessInsider.com, 29 July 2013, www.BusinessInsider.com/youversion-bible-app-has-100-million-downloads-2013-7.

7 Tomi Ahonen, 'The Annual Industry Numbers and Stats Blog – Yep, this year we will hit the mobile moment', blog post on Communities Dominate.blogs.com, 6 March 2013, communitiesdominate.blogs.com/ brands/2013/03/the-annual-mobile-industry-numbers-and-stats-blog-yep-this-year-we-will-hit-the-mobile-moment.html.

8 Mayumi Negishi, 'Rakuten to Buy Voice-Call App Maker Viber', article on WSJ.com, 14 February 2014, online.wsj.com/news/articles/ SB10001424052702304315004579382014046629596.

9 Sven Grundberg and Juhana Rossi, 'Earnings Soar at Finnish Game Maker Supercell', article on WSJ.com, 12 February 2014, online.wsj.com/ news/articles/sb10001424052702304703804579378272705325260?mg=reno 4-wsj.

10 Mary Meeker and Liang Wu, KPCB Annual Internet Trends Report 2013, published May 2013, slide 14, www.kpcb.com/insights/2013-internet-trends.

11 Spinvox, profile on CrunchBase.com, www.crunchbase.com/company/ spinvox.

12 'A Voicemail Transcription Scandal in Britain', blog post on WSJ.com, 24 July 2009, blogs.wsj.com/digits/2009/07/24/a-voicemail-transcription-scandal-in-britain/.

13 Andrew Orlowski, 'Spinvox: The Inside Story', article on TheRegister.co.uk, 28 July 2009, www.theregister.co.uk/2009/07/29/spinvox_mechanical_turk/.

14 Diana I. Tamir and Jason P. Mitchell, 'Disclosing Information About the Self is Intrinsically Rewarding', *PNAS*, volume 109, number 21, 22 May 2012, wjh.harvard.edu/~dtamir/Tamir-PNAS-2012.pdf.

15 Mary Meeker and Liang Wu, May 2013, slide 28, op. cit.

16 Reed Albergotti, Douglas MacMillan and Evelyn M. Rusli, 'Facebook to Pay $19 Billion for WhatsApp', article on WSJ.com, 19 February 2014, online.wsj.com/news/articles/SB100014240527023049142045793934520 9288302.

17 Liz Gannes, 'WhatsApp CEO Jan Koum Hates Advertising and the Tech Rumor Mill (Full Dive Video)', article on AllThingsD.com, 10 May 2013, allthingsd.com/20130510/whatsapp-ceo-jan-koum-hates-advertising-and-the-tech-rumor-mill-full-dive-video/.

18 Reed Albergotti, Douglas MacMillan and Evelyn M. Rusli, 19 February 2014, op. cit.

19 George Anders, 'Facebook's $19 Billion Craving, Explained by Mark Zuckerberg', article on Forbes.com, 19 February 2014, www.forbes.com/sites/georgeanders/2014/02/19/facebook-justifies-19-billion-by-awe-at-whatsapp-growth/.

20 Interview with Jan Koum at the DLD (Digital-Life-Design) Conference 2014, moderated by David Rowan, video published on YouTube, 20 January 2014, www.youtube.com/watch?v=WgAtBTpm6Xk.

21 David Meyer, 'Chat Apps Have Overtaken SMS by Message Volume, But How Big a Disaster is That for Carriers?', article on GIGAOM.com, 29 April 2013, gigaom.com/2013/04/29/chat-apps-have-overtaken-sms-by-message-volume/.

22 Interview with Jan Koum, 20 January 2014, op. cit.

23 J. J. Colao, 'Snapchat: The Biggest No-Revenue Mobile App Since Instagram', article on Forbes.com, 27 November 2012, www.forbes.com/sites/jjcolao/2012/11/27/snapchat-the-biggest-no-revenue-mobile-app-since-instagram/.

24 Evelyn M. Rusli and Douglas MacMillan, 13 November 2013, op. cit.

25 Mike Isaac, 'Snapchat Closes $60 Million Round Led by IVP, Now at 200 Million Daily Snaps', article on AllThingsD.com, 24 June 2013, allthingsd.com/20130624/snapchat-closes-60-million-round-led-by-ivp-now-at-200-million-daily-snaps/.

26 'Recent Additions to Team Snapchat', blog post on Snapchat.com, 24 June 2013, blog.snapchat.com/post/53763657196/recent-additions-to-team-snapchat.

27 Mike Isaac, 'Snapchat Now Boasts More Than 150 Million Photos Taken Daily', article on AllThingsD.com, 16 April 2013, allthingsd.com/20130

416/snapchat-now-boasts-more-than-150-million-photos-taken-daily/.

28 Justin Lafferty, 'Facebook Photo Storage Is No Easy Task', article on AllFacebook.com, 16 January 2013, allfacebook.com/facebook-photo-storage-open-compute_b108640.

Part II: The Journey

Chapter 4: It's Bloody Hard

1 Aileen Lee, 'Welcome To The Unicorn Club: Learning From Billion-Dollar Startups', article on TechCrunch.com, 2 November 2013, TechCrunch.com/2013/11/02/welcome-to-the-unicorn-club/.

2 http://graphics.wsj.com/billion-dollar-club-the.

3 According to NVCA figures reported in Aileen Lee's article 'Welcome To The Unicorn Club: Learning From Billion-Dollar Startups' published on TechCrunch.com, 16,000 Internet-related companies were funded since 2003; Mattermark says that 12,291 were funded in the past 2 years, and CVR says about 10–15,000 software companies are seeded every year. This makes about 60,000 in the last decade.

4 Mark Lennon, 'CrunchBase Reveals: The Average Successful Startup Raises $41M, Exits at $242.9M', article on TechCrunch.com, 14 December 2013, TechCrunch.com/2013/12/14/crunchbase-reveals-the-average-successful-startup-raises-41m-exits-at-242-9m/.

5 According to separate studies by the US Bureau of Labor Statistics and the Ewing Marion Kauffman Foundation.

6 Deborah Gage, 'The Venture Capital Secret: 3 Out of 4 Start-Ups Fail', article on WSJ.com, 20 September 2012, online.wsj.com/news/articles/SB10000872396390443720204578004980476429190.

7 www.crunchbase.com.

8 Aileen Lee, 2 November 2013, op. cit.

9 Ibid.

10 Ibid.

11 Ibid.

12 Evelyn M. Rusli, 'Instagram Pictures Itself Making Money', article on WSJ.com, 8 September 2013, online.wsj.com/news/articles/SB10001424127887324577304579059230069305894.

13 'First Look – Measuring the Effectiveness of Brand Advertising on Instagram', blog post on Instagram-Business.tumblr.com, December 2013, instagram-business.tumblr.com/post/70498340316/first-look-measuring-the-effectiveness-of-brand.

14 Kara Swisher, 'The Money Shot', article for *Vanity Fair*, June 2013, www.vanityfair.com/business/2013/06/kara-swisher-instagram.

15 Information taken from the 'About Us' section on Instagram.com, instagram.com/about/us/.

16 Kara Swisher, June 2013, op. cit.

17 Ibid.

18 Shayndi Raice, Spencer E. Ante and Emily Glazer, 'In Facebook Deal, Board
    Was All But Out of Picture', article on WSJ.com, 18 April 2012, online.
    wsj.com/news/articles/SB10001424052702304818404577350191931921290.

19 Kara Swisher, June 2013, op. cit.

Step 1: The Million-Dollar App

1 http://pitchbook.com/1Q_2014.html.

Chapter 5: Let's Get Started

1 Aileen Lee, 'Welcome To The Unicorn Club: Learning From Billion-Dollar
    Startups', article on TechCrunch.com, 2 November 2013, TechCrunch.
    com/2013/11/02/welcome-to-the-unicorn-club/.

2 Jacob Mullins, 'Want To Build A $1B Consumer Company? Look for
    Long-Haul Founders and Don't Fear Incumbents', article on
    TechCrunch.com, 2 March 2013, TechCrunch.com/2013/03/02/how-do-
    you-build-a-1b-consumer-company/.

3 Aileen Lee, 2 November 2013, op. cit.

4 Ibid.

5 http://www.census.gov/prod/2005pubs/p70-97.pdf.

6 Aileen Lee, 2 November 2013, op. cit.

7 To find out more about the TechCrunch Disrupt Hackathon, visit
    www.hackerleague.org/hackathons/techcrunch-disrupt-sf-2013.

8 Anthony Ha, 'Tristan O'Tierney, Square's Co-Founder and Early iOS
    Engineer, Leaves for Destinations Unknown', article on TechCrunch.com,
    15 June 2013, TechCrunch.com/2013/06/15/tristan-tierney-leaves-square/.

9 Jason Pontin, 'Who Owns the Concept if No One Signs the Papers?',
    article on NYTimes.com, 12 August 2007, www.nytimes.com/
    2007/08/12/business/yourmoney/12stream.html?_r=2&.

10 Richard MacManus, 'How Flipboard Was Created and its Plans Beyond
    iPad', article on ReadWrite.com, 6 October 2010, readwrite.com/
    2010/10/06/how_flipboard_was_created_its_plans_beyond_ipad.

Chapter 6: Solving the Identity Crisis

1 Fred Wilson, 'Finding and Buying a Domain Name', blog post on
    AVC.com, 28 April 2011, www.AVC.com/a_vc/2011/04/finding-and-
    buying-a-domain-name.html.

2 Sarah Lacy, 'Get Over It, Haters: 99designs Had Tipped', article on
    Pando.com, 24 January 2012, pando.com/2012/01/24/get-over-it-haters-
    99designs-has-tipped/.

Chapter 7: Getting Lean and Mean

1 Tiurivan Agten, 'A Granular App Level Look at Revenues: Google Play vs. Apple App Store', blog post on Distimo.com, 29 May 2013, www.distimo.com/blog/2013_05_a-granular-app-level-look-at-revenues-google-play-vs-apple-app-store/.

2 Ben Parr, '"Angry Birds" Hits 42 Million Free and Paid Downloads', article on Mashable.com, 9 December 2010, mashable.com/2010/12/08/angry-birds-hits-42-million-downloads/.

3 Mike Rose, 'Supercell's Secret Sauce', article on Gamasutra.com, www.gamasutra.com/view/feature/183064/supercells_secret_sauce.php? print=1.

4 'Apple picks Flipboard as App of the Year!', blog post on Inside.Flipboard.com, 9 December 2010, inside.flipboard.com/2010/12/09/apple-picks-flipboard-as-app-of-the-year/.

5 'Gartner Says Worldwide Traditional PC, Tablet, Ultramobile and Mobile Phone Shipments On Pace to Grow 7.6 Percent in 2014', article on Gartner.com, 7 January 2014, www.gartner.com/newsroom/id/2645115.

6 'Android Fragmentation Visualized', report on OpenSignal.com, August 2012, opensignal.com/reports/fragmentation.php.

7 Juli Clover, 'iOS 7 Now on 73% of Devices, but Adoption Rates "Much Slower" Than iOS 6', article on MacRumors.com, 18 October 2013, www.macrumors.com/2013/10/18/ios-7-now-on-73-of-devices-but-adoption-rates-much-slower-than-ios-6/.

8 'Android Fragmentation Visualized', August 2012, op. cit.

Chapter 8: App Version 0.1

1 Anthony Wing Kosner, 'Jony Ives' (No Longer So) Secret Design Weapon', article on Forbes.com, 30 November 2013, www.forbes.com/sites/anthonykosner/2013/11/30/jony-ives-no-longer-so-secret-design-weapon/.

2 'SFMOMA Presents Less and More: The Design Ethos of Dieter Rams', press release, 29 June 2011, www.sfmoma.org/about/press/press_exhibitions/releases/880.

3 Brian Suthoff, 'First Impressions Matter! 26% of Apps Downloaded in 2010 Were Used Just Once', blog post on Localytics.com, 31 January 2011, www.localytics.com/blog/2011/first-impressions-matter-26-percent-of-apps-downloaded-used-just-once/.

Chapter 10: Let's Get Some Users

1 http://www.slideshare.net/victori98pt/state-of-mobile-q32011-by-nielsen-per-cent20.

2 Natasha Lomas, 'Mainline App Stores Still Dominate iOS/Android App Discovery, Finds Forrester, But Word of Mouth & Social Recommendations

Also Key', article on TechCrunch.com, 17 April 2013, TechCrunch.com/
2013/04/17/forrester-app-discovery-report/.

3 Scott Reyburn, 'Google Discloses How Search For Google Play Works for
the First Time; 12 percent of DAU search for apps daily', article on
InsideMobileApps.com, 17 May 2013, www.insidemobileapps.com/
2013/05/17/google-discloses-how-search-for-google-play-works-for-the-
first-time-12-percent-of-dau-search-for-apps-daily/.

4 List compiled from information provided on Moz.com. To find out more,
visit moz.com/beginners-guide-to-seo.

5 Link farming is the 'process of exchanging reciprocal links with Web sites
in order to increase search engine optimisation'. See www.webopedia.com/
TERM/L/link_farming.html.

6 Matt Marshall, 'The Top 10 Mobile Advertising Companies', article on
VentureBeat.com, 12 June 2013, venturebeat.com/2013/06/12/the-top-
10-mobile-advertising-companies/.

7 'What Does it Cost to do Press Releases and What Services (Marketwire,
PRWeb, Business Wire) Are Best?', Quora.com, www.quora.com/What-
does-it-cost-to-do-press-releases-and-what-services-Marketwire-PRWeb-
BusinessWire-etc-are-best.

Chapter 11: Is Your App Ready for Investment?

1 According to a post from a partner at Y Combinator on 'How Many
People/Teams Get Rejected by Y Combinator During Each Application
Period?', Quora.com, www.quora.com/How-many-people-teams-get-
rejected-by-Y-Combinator-during-each-application-period.

2 This information comes from an interview with Alice Bentinck, cofounder
of Entrepreneur First conducted on 24 February 2014.

Chapter 12: How Much is Your App Worth and How Much Money Should You Raise?

1 Bill Payne, 'The Pre-Money Valuation of Angel Deals in 2012', blog post on
AngelCapitalAssociation.org, 13 May 2013, www.angelcapitalassociation.
org/blog/1266679842/.

2 Investor Tool on SiSense.com, www.crunchbase.sisense.com/
#!investor-tool/component_74511.

3 When you get to the point of issuing shares it is definitely worth con-
sulting an accountant or tax adviser to make sure that you are complying
with any relevant tax rules in your country. You should think about the
percentage ownership carefully as it may cause tax problems to try and
transfer shares around among cofounders later on.

4 Profile of John Doerr in the *Forbes* 400, www.forbes.com/profile/
john-doerr/.

5 To view the documents, visit www.techstars.com/docs/.

6 To view the New Plain Preferred Term Sheet visit fi.co/posts/69.

Step 2: The Ten-Million-Dollar App

Chapter 14: Make Something People Love

1 'The Pmarca Guide to Startups, part 4: The only thing that matters', blog post on blog.Pmarca.com, 25 June 2007, web.archive.org/web/20070701074943; http://blog.pmarca.com/2007/06/the-pmarca- gu-2.html.
2 Sean Ellis, 'The Startup Pyramid', article for Startup-Marketing.com, http://www.startup-marketing.com/the-startup-pyramid/.
3 Chamath Palihapitiya, 'How We Put Facebook on the Path to 1 billion Users', part of a 10-hour course from the Growth Hackers Conference, published 9 January 2013, www.youtube.com/watch?v=raIUQP71SBU.

Chapter 16: The Metrics of Success

1 Paul Graham, 'How to Start a Startup', blog post on PaulGraham.com, March 2005, paulgraham.com/start.html.

Chapter 17: Getting Your Growth On

1 For more information about Mobile App Tracking, visit mobileapptracking.com/.
2 'An Introduction to Mobile App Tracking', 1 May 2012, www.slideshare.net/MobileAppTracking/mobile-app-tracking-how-it-works.

Chapter 18: Dollars in the Door

1 Greg Kumparak, 'Want To Raise A Million Bucks? Here's What You'll Need', article on TechCrunch.com, 22 April 2013, TechCrunch.com/2013/04/22/want-to-raise-a-million-bucks-heres-what-youll-need/.
2 Dave McClure, 'Changes in Tech Startups and Venture Capital', presentation to the Asian Business Angel Forum, slide 11, www.slideshare.net/dmc500hats/changes-in-venture-capital-tech-startups-nov-2013-mumbai.

Chapter 19: Seducing Venture Capital

1 Allen Wagner, 'Increasing Seed Valuations Exacerbate "Series A Crunch"', blog post on PitchBook.com, 15 October 2013, blog.pitchbook.com/increasing-seed-valuations-exacerbate-series-a-crunch/.
2 The Entrepreneurs Report: Private Company Financing Trends, 'Financing Trends for Q3 2012', www.wsgr.com/publications/PDFSearch/entreport/Q32012/private-company-financing-trends.htm.
3 Allen Wagner, 15 October 2013, op. cit.
4 'Series B May Be the Real Startup Funding Crunch', article on BizJournals.com, 14 May 2013, www.bizjournals.com/sanjose/news/2013/03/14/series-b-may-be-the-real-startup.html?s=image_gallery.

5 Brad Feld, 'VC Rights: Up, Down, and Know What The Fuck Is Going On', blog post on Feld.com, 7 May 2012, www.feld.com/wp/archives/category/term-sheet.
6 Ibid.
7 Fred Wilson, 'The Three Terms You Must Have in a Venture Investment', blog post on AVC.com, 10 April 2009, www.AVC.com/a_vc/2009/04/the-three-terms-you-must-have-in-a-venture-investment.html.

Step 3: The Hundred-Million-Dollar App

1 http://pitchbook.com/1Q_2014.html.

Chapter 20: A Colorful Lesson

1 Claire Cain Miller, 'Investors Provide Millions to Risky Startups', article on NYTimes.com, 19 June 2011, www.nytimes.com/2011/06/20/technology/20color.html.
2 'Bill Nguyen: The Boy in the Bubble', article on FastCompany.com, 19 October 2011, www.fastcompany.com/1784823/bill-nguyen-boy-bubble.

Chapter 21: Tuning and Humming

1 Shopzilla, profile on CrunchBase.com, www.crunchbase.com/company/shopzilla.
2 'Marketer's #1 Concern: Mobile Tracking and Measurement – Way Beyond Cost, Privacy and Scaling', blog post on AdTruth.com, 21 September 2012, blog.adtruth.com/marketers-1-concern-mobile-tracking-and-measurement-way-beyond-cost-privacy-and-scaling/.
3 Fred Wilson, 'Employee Equity: How Much?', blog post on AVC.com, www.AVC.com/a_vc/2010/11/employee-equity-how-much.html.

Chapter 22: Getting Shedloads of Users

1 Heather Leonard, 'BII Mobile Insights: The Top Mobile Predictions for 2013', article on BusinessInsider.com, 12 December 2012, www.BusinessInsider.com/bii-mobile-insights-top-mobile-predictions-for-2013-2012-12.
2 If you want to read more about Fiksu, see its e-books which are available free online at www.fiksu.com/resources/ebooks.
3 Figure taken from information provided on www.addshoppers.com/stats/.
4 Matt Anderson, Joe Sims, Jerell Price and Jennifer Brusa, 'Turning "Like" to "Buy" Social Media Emerges as a Commerce Channel', article on Booz.com, www.booz.com/media/uploads/BaC-Turning_Like_to_Buy.pdf.

5 Cooper Smith and Marcelo Ballve, '5 Charts That Show How Mobile and Social Media Are Taking Over Commerce', article on BusinessInsider.com, 22 October 2013, www.BusinessInsider.com/social-and-mobile-are-eating-retail-2013-10.

6 Ingrid Lunden, 'Digital Ads Will Be 22% of All US Ad Spend in 2013, Mobile Ads 3.7%; Total Global Ad Spend in 2013 $503B', article on TechCrunch.com, 30 September 2013, TechCrunch.com/2013/09/30/digital-ads-will-be-22-of-all-u-s-ad-spend-in-2013-mobile-ads-3-7-total-gobal-ad-spend-in-2013-503b-says-zenithoptimedia/.

7 Ibid.

8 Alex Cocotas, 'The Social Media Advertising Ecosystem Explained', article on BusinessInsider.com, 24 August 2013, www.BusinessInsider.com/state-of-social-media-advertising-2013-7.

9 Josh Constine, 'Facebook Reveals 78% of US Users Are Mobile as It Starts Sharing User Counts by Country', article on TechCrunch.com, 13 August 2013, TechCrunch.com/2013/08/13/facebook-mobile-user-count/.

10 'Mobile User Bases at Facebook and Twitter Keep Growing', article on eMarketer.com, 15 August 2013, www.emarketer.com/Article/Mobile-User-Bases-Facebook-Twitter-Keep-Growing/1010135l.

11 Cheryl Morris, 'Global Facebook Advertising KPIs – Retail Benchmark Report, Part 1', article on Nanigans.com, 17 October 2013, www.nanigans.com/2013/10/17/global-facebook-advertising-kpis-retail-benchmark-report-part-i/.

12 John Koetsier, 'Facebook Ad Profit a Staggering 1,790% More on iPhone than Android', article on VentureBeat.com, 16 October 2013, venturebeat.com/2013/10/16/facebook-ad-profit-a-staggering-1790-more-on-iphone-than-android/.

Chapter 23: Revenue-Engine Mechanics

1 David Skok, 'Startup Killer: The Cost of Customer Acquisitions', article on forEntrepreneurs.com, 22 December 2009, www.forentrepreneurs.com/startup-killer/.

2 Eliana Dockterman, 'Candy Crush Saga: The Science Behind Our Addiction', article on Time.com, 15 November 2013, business.time.com/2013/11/15/candy-crush-saga-the-science-behind-our-addiction/.

3 Mia Shanley, 'How Candy Crush Makes So Much Money', article on BusinessInsider.com, 8 October 2013, www.BusinessInsider.com/how-candy-crush-makes-so-much-money-2013-10.

4 Dave McClure, 'Startup Metrics for Pirates', presented to Wildfire Interactive, May 2012, slide 73, www.slideshare.net/dmc500hats/startup-metrics-4-pirates-wildfire-interactive-may-2012.

5 Mike Isaac, 'Snapchat Closes $60 Million Round Led by IVP, Now at 200 Million Daily Snaps', article on AllThingsD.com, 24 June 2013,

allthingsd.com/20130624/snapchat-closes-60-million-round-led-by-ivp-now-at-200-million-daily-snaps/.

6 Mike Isaac, 'Snapchat Now Boasts More Than 150 Million Photos Taken Daily', article on AllThingsD.com, 16 April 2013, allthingsd.com/20130416/snapchat-now-boasts-more-than-150-million-photos-taken-daily/.

7 Liz Gannes, 'Popular Photo Message App Snapchat Adds Video', article on AllThingsD.com, 14 December 2012, allthingsd.com/20121214/popular-photo-message-app-snapchat-adds-video/.

8 David Skok, 'The Science Behind Viral Marketing', article on forEntrepreneurs.com, 15 September 2011, www.forentrepreneurs.com/the-science-behind-viral-marketing/.

Chapter 24: Keeping Users Coming Back

1 Fred Wilson, '30/10/10', blog post on AVC.com, 30 July 2011, www.AVC.com/a_vc/2011/07/301010.html.

2 'Trends Report: The New Standards for Mobile App Retention', blog post on Mixpanel.com, 4 November 2013, blog.mixpanel.com/2013/11/04/trends-report-the-new-standards-for-mobile-app-retention/.

3 Ibid.

4 To find out more about Mixpanel, visit www.mixpanel.com.

Chapter 25: International Growth

1 Travis Kalanick, CEO of Uber Technologies, speaking at Fortune's 2013 Brainstorm Tech Conference, video published on 23 July 2013, www.youtube.com/watch?v=vGbuitwkZiM.

2 Christine Lagorio-Chafkin, 'Resistance Is Futile', article for *Inc.* magazine, July/August 2013 issue, www.inc.com/magazine/201307/christine-lagorio/uber-the-car-service-explosive-growth.html.

3 Eric Schneiderman, 'Airbnb Hit with Subpoena from NY Attorney General', blog post on WSJ.com, 7 October 2013, blogs.wsj.com/metropolis/2013/10/07/airbnb-hit-with-subpoena-from-n-y-attorney-general/.

4 Dan Primack, 'More details on Uber's Massive Funding Round', article on CNN.com, 23 August 2013, finance.fortune.cnn.com/2013/08/23/more-details-on-ubers-massive-funding-round/.

5 Catherine Shu, 'Square Starts Mobile Payments in Japan, Its First Country Outside of North America, In Partnership With Visa's Ally', article on TechCrunch.com, 23 May 2013, TechCrunch.com/2013/05/23/square-starts-mobile-payments-in-japan-its-first-country-outside-of-north-america-in-partnership-with-visas-ally/.

6 Information taken from squareup.com/jp.

7 Jeff Blagdon, 'Square Arrives in Japan, Its First Market Outside North America', article on TheVerge.com, 23 May 2013, www.theverge.com/

2013/5/23/4358294/jack-dorsey-square-tokyo-japan.

8 Grace Huang and Takashi Amano, 'Apple Won 76% of Japan October Smartphone Sales, Kantar Says', article on Bloomberg.com, 29 November 2013, www.bloomberg.com/news/2013-11-28/apple-won-76-of-japan-smartphone-sales-in-october-kantar-says.html.

9 Hugo Barra interview with LeWeb in Paris, video published on 11 December 2013, www.youtube.com/watch?v=mZsvJUa9FpI.

10 Ibid.

11 Data retrieved from www.cnnic.net.cn/hlwfzyj/hlwxzbg/hlwtjbg/201401/P020140116395418429515.pdf .

12 Evelyn M. Rusli and Douglas MacMillan, 'Snapchat Spurned $3 Billion Acquisitions Offer from Facebook', blog post on WSJ.com, 13 November 2013, blogs.wsj.com/digits/2013/11/13/snapchat-spurned-3-billion-acquisition-offer-from-facebook/.

Chapter 27: Money for Scale

1 'Average Times Between Venture Capital rounds', SiSense.com, www.crunchbase.sisense.com/#!latest-tech-trends/cy9t.

2 http://pitchbook.com/1Q_2014.html.

3 Alexia Tsotsis, 'Snapchat Snaps Up a $80M Series B Led by IVP at an $800 Million Valuation', article on TechCrunch.com, 22 June 2013, TechCrunch.com/2013/06/22/source-snapchat-snaps-up-80m-from-ivp-at-a-800m-valuation/.

4 Alexia Tsotsis, 'Uber Hires European Kees Koolen at COO, To Help It Scale Globally', article on TechCrunch.com, 19 June 2012, TechCrunch.com/2012/06/19/uber-hires-a-european-coo-to-help-it-scale-globally/.

5 Ingrid Lunden, 'Angry Birds Maker Rovio Says 2012 Sales Up 101% To $195M with Merchandising, IP 45% of That; Net Profit $17M', article on TechCrunch.com, 3 April 2013, TechCrunch.com/2013/04/03/rovios-revenues-up-101-to-195m-non-games-45-of-that-net-profit-71m/.

6 http://pitchbook.com/1Q_2014.html.

7 Fred Wilson, 'Financing Options: Bridge Loans', blog post on AVC.com, 15 August 2011, www.AVC.com/a_vc/2011/08/financing-options-bridge-loans.html.

8 Ibid.

Step 4: The Five-Hundred-Million-Dollar App

Chapter 28: Shifting Up a Gear

1 Alex Taussig, 'How to Know if Your Business Will Scale', article on CNN.com, 1 June 2011, finance.fortune.cnn.com/2011/06/01/but-does-it-scale/.

2 J. J. Colao, 'The Inside Story of Snapchat: The World's Hottest App or A $3 Billion Disappearing Act?', article for *Forbes*, 20 January 2014, www.forbes.com/sites/jjcolao/2014/01/06/the-inside-story-of-snapchat-the-worlds-hottest-app-or-a-3-billion-disappearing-act/.

Chapter 29: Big Hitters

1 'Uber's New COO and Partner', blog post on Uber.com, 19 June 2012, blog.uber.com/2012/06/19/ubers-new-coo-and-partner/.
2 Information taken from Kees Koolan's LinkedIn profile, www.linkedin.com/in/keeskoolen.
3 Kara Swisher, 'Instagram Business Lead Emily White to Be Named COO of Snapchat', article on AllThingsD.com, 3 December 2013, allthingsd.com/20131203/exclusive-instagram-business-lead-emily-white-to-be-named-coo-of-snapchat/.
4 Information taken from Emily White's LinkedIn profile, www.linkedin.com/pub/emily-white/7/767/20b.

Chapter 30: Scaling Marketing

1 'What's Your Exit Strategy This Season?', press release on Hailocab.com, 27 November 2013, hailocab.com/toronto/press-releases/Whats-your-Exit-Strategy-this-Holiday-Season.
2 Drew Olanoff, 'Square Wallet Now Accepted in Over 7,000 Starbucks Stores in the US, Only 3 Months After Deal Signed', article on TechCrunch.com, 7 November 2012, TechCrunch.com/2012/11/07/square-wallet-will-now-be-accepted-in-over-7000-starbucks-stores-in-the-united-states/.
3 Stuart Dredge, 'Candy Crush Saga: "70% of the people on the last level haven't paid anything"', article on TheGuardian.com, 10 September 2013, www.theguardian.com/technology/appsblog/2013/sep/10/candy-crush-saga-king-interview.

Chapter 31: Killer Product Expansion

1 Mark Walsh, 'Flipboard CEO Explains How Brand Is Monetizing Users', article on MediaPost.com, 7 October 2013, www.mediapost.com/publications/article/210749/flipboard-ceo-explains-how-brand-is-monetizing-users.html.
2 Ibid.
3 Austin Carr, 'With 90M Users, Flipboard Launches Shopping Magazine Experience', article on FastCompany.com, 11 November 2013, www.fastcompany.com/3021443/with-90m-users-flipboard-launches-shopping-magazine-experience.
4 Matt Rosoff, 'Square COO Keith Rabois Explains the Death of the Web and How Square Will Move Beyond Payments', article on BusinessInsider.com,

6 April 2012, www.BusinessInsider.com/keith-rabois-qa-with-the-square-coo-and-angel-investor-2012-4.

5 Seth Fieherman, 'Rovio Generated $106 Million From Angry Birds in 2011', article on BusinessInsider.com, 7 May 2012, www.BusinessInsider.com/rovio-made-106-million-from-angry-birds-in-2011-2012-5.

6 Kevin Smith, 'Angry Birds Maker Rovio Reports $200 Million in Revenue, $71 Million in Profit for 2012', article on BusinessInsider.com, 3 April 2013, www.BusinessInsider.com/angry-birds-made-200-million-in-2012-2013-4.

7 'Rovio Entertaining Reports 2012 Financial Results', press release on Rovio.com, 3 April 2013, www.rovio.com/en/mobile-news/284/rovio-entertainment-reports-2012-financial-results.

Chapter 32: Scaling Product Development and Engineering

1 Billy Gallagher, 'Snapchat Now Sees 350M Photos Shared Daily, Up From 200M in June', article on TechCrunch.com, 9 September 2013, TechCrunch.com/2013/09/09/snapchat-now-sees-350m-photos-shared-daily-up-from-200m-in-june/.

2 WhatsApp revealed these figures through the company's official Twitter account @WhatsApp on 2 January 2013, twitter.com/WhatsApp/status/286591302185938946.

3 Alyson Shontell, 'The 25 Best-Paying Companies for Software Engineers', article on BusinessInsider.com, 2 April 2013, www.BusinessInsider.com/the-worlds-highest-paid-software-engineers-work-for-these-25-companies-2013-4.

4 Bureau of Labor Statistics, US Department of Labor, *Occupational Outlook Handbook, 2014-15 Edition*, Software Developers, published 8 January 2014, http://www.bls.gov/ooh/computer-and-information-technology/software-developers.htm.

5 Bureau of Labor Statistics, US Department of Labor, 'Occupational Employment and Wages, May 2012', published on bls.gov, 29 March 2013, www.bls.gov/oes/current/oes271024.htm.

6 Tim Peterson, 'Facebook Picks Up Design Agency Hot Studio: Acqui-hire Brings First Designers to Company's New York Office', article on AdWeek.com, 14 March 2013, www.adweek.com/news/advertising-branding/facebook-picks-design-agency-hot-studio-147942.

7 M.G. Seigler, 'Facebook Buys Sofa, a Software Design Team That Will Help Make Facebook More Beautiful', article on TechCrunch.com, 9 June 2011, TechCrunch.com/2011/06/09/facebook-sofa/.

8 Josh Constine, 'Facebook Acq-Hires Part of Design Firm Bolt I Peters to Beef Up User Research Team', article on TechCrunch.com, 24 May 2012, TechCrunch.com/2012/05/24/facebook-bolt-peters/.

9 Claire Cain Miller, 'Square Acquires a New York Design Firm', blog post

on NYTimes.com, 1 October 2012, bits.blogs.nytimes.com/2012/10/01/square-acquires-a-new-york-design-firm/?_r=1.

Chapter 33: Scaling People

1 James Altucher, 'Why You Shouldn't Act Like the Billionaires You Respect', article on BusinessInsider.com, 26 April 2013, www. Business Insider.com/dont-act-like-a-billionaire-2013-4.

2 Matt Rosoff, 'Square COO Keith Rabois Explains the Death of the Web and How Square Will Move Beyond Payments', article on Business Insider.com, 6 April 2012, www.BusinessInsider.com/keith-rabois-qa-with-the-square-coo-and-angel-investor-2012-4.

3 Jason Del Rey, 'Square Loses Two Execs, Including One Out Before His First Day on the Job', article on AllThingsD.com, 22 May 2013, allthingsd.com/20130522/square-loses-two-execs-including-one-out-before-his-first-day-on-the-job/?mod=atdtweet.

Chapter 34: Scaling Process

1 Version One Ventures, 'How to Stay Nimble as You Scale', article on VersionOneVentures.com, 30 September 2013, versiononeventures.com/stay-nimble-scale/.

2 Robert Medrano, 'Welcome to the API Economy', guest post on Forbes.com, 29 August 2012, www.forbes.com/sites/ciocentral/2012/08/29/welcome-to-the-api-economy/.

3 Version One Ventures, 30 September 2013, op. cit.

4 The letter can be viewed at www.sec.gov/Archives/edgar/data/1326801/000119312512034517/d287954ds1.htm.

5 Paul Graham, 'Maker's Schedule, Manager's Schedule', blog post on PaulGraham.com, July 2009, www.paulgraham.com/makersschedule.html.

6 Ibid.

7 Fred Wilson, 'Tech Ops as a Metaphor for Building, Running and Leading a Company', blog post on AVC.com, 15 October 2013, www.AVC.com/a_vc/2013/10/tech-ops-as-a-metaphor-for-building-running-leading-a-company.html.

8 Version One Ventures, 30 September 2013, op. cit.

9 Kristen Gil, 'Start-Up Speed', article on ThinkWithGoogle.com, January 2012, www.google.com/think/articles/start-up-speed-kristen-gil.html.

10 Larry Page and Q&A with Eric Schmidt at Zeitgeist Americas 2011, video uploaded to YouTube on 27 September 2011, www.youtube.com/watch?v=srI6QYfi-HY.

Chapter 35: Financing at a Big Valuation

1 Kara Swisher, 'Uber Filing in Delaware Shows TPG Investment at $3.5 Billion Valuation: Google Ventures Also In', article on AllThingsD.com, 22 August 2013, allthingsd.com/20130822/uber-filing-in-delaware-shows-tpg-investment-at-3-5-billion-valuation-google-ventures-also-in/.
2 Ibid.
3 'The B Word – What Does It Take to Raise a Financing Round at a Billion Dollar Valuation?', article on CBInsights.com, 7 February 2014, www.cbinsights.com/blog/trends/billion-dollar-startups-venture-capital.
4 Ibid.

Step 5: The Billion-Dollar App: The Promised Land

Chapter 36: Unicorns Do Exist

1 Information posted on Brian Acton's Twitter account, @brianacton, on 23 May 2009, twitter.com/brianacton/statuses/1895942068.
2 Information posted on Brian Acton's Twitter account, @brianacton, on 3 August 2009, twitter.com/brianacton/status/3109544383.

Chapter 37: People at a Billion-Dollar Scale

1 Victoria Baret, 'How the Kids at Box Are Disrupting Software's Most Lucrative Game', *Forbes*, 4 March 2013 issue, www.forbes.com/sites/victoriabarret/2013/02/13/box-aaron-levie-mobile-enterprise-software/.
2 Eric Savitz, 'SurveyMonkey To Raise $794M In Recap; Valuation $1.35 Billion (Updated)', article on Forbes.com, 17 January 2013, www.forbes.com/sites/ericsavitz/2013/01/17/surveymonkey-to-raise-794m-in-recap-valuation-1-35-billion/.
3 'How Dave Goldberg of SurveyMonkey Built a Billion-Dollar Business and Still Gets Home by 5.30 p.m.', article and video interview on FirstRound.com, firstround.com/article/how-dave-goldberg-of-survey-monkey-built-a-billion-dollar-business-and-still-gets-home-by-5-30.
4 Ibid.
5 Mike Rose, 'Supercell's Secret Sauce', article on Gamasutra.com, 7 December 2012, www.gamasutra.com/view/feature/183064/supercells_secret_sauce.php.
6 Ibid.
7 Alyson Shontell and Andrea Huspeni, '15 Incredible Employee Perks That Will Make You Wish You Worked at a Startup', article on BusinessInsider. com, 31 May 2012, www.BusinessInsider.com/killer-startup-perks-2012-5.
8 Heather Leonard, 'Facebook Generates Over $1 Million in Revenue Per Employee', article on BusinessInsider.com, 19 March 2013, www.BusinessInsider.com/facebook-has-high-revenue-per-employee-2013-3.

9 Megan Rose Dickey, '"Clash of Clans" Maker Had a Monster Year in 2013: Revenue Increased Nearly Ninefold', article on BusinessInsider.com, 12 February 2014, www.BusinessInsider.com/gaming-startup-supercell-2013-revenue-2014-2.

10 Steven Levy, 'Google's Larry Page on Why Moon Shots Matter', article on Wired.com, 17 January 2013, www.wired.com/business/2013/01/ff-qa-larry-page/all/.

11 Peter Murray, 'Google's Self-Driving Car Passes 300,000 Miles', article on Forbes.com, 15 August 2012, www.forbes.com/sites/singularity/2012/08/15/googles-self-driving-car-passes-300000-miles/.

12 For more information about Project Loon, visit www.google.com/loon/.

13 'Google X', entry on Wikipedia, en.wikipedia.org/wiki/Google_X.

Chapter 38: Advice from Billion-Dollar CEOs

1 Will Oremus, 'Google's Big Break', article on Slate.com, 13 October 2013, www.slate.com/articles/business/when_big_businesses_were_small/2013/10/google_s_big_break_how_bill_gross_goto_com_inspired_the_adwords_business.html.

2 Ibid.

3 'Drew Houston's Morph from Hacker to Hyper-Growth CEO', article on FirstRound.com, www.firstround.com/article/Drew-Houstons-morph-from-hacker-to-hyper-growth-CEO.

4 Peter Kafka, 'Larry Page on Speed: "There are no companies that have good slow decisions"', article on AllThingsD.com, 27 September 2011, allthingsd.com/20110927/larry-page-on-speed-there-are-no-companies-that-have-good-slow-decisions/.

5 Glen Cathey, 'LinkedIn Traffic Statistics and User Demographics 2013', article on BooleanBlackBelt.com, 24 July 2013, booleanblackbelt.com/2013/07/linkedin-traffic-statistics-and-user-demographics-2013/.

6 Juhana Hietala, 'Rovio Mobile Company Presentation – Dynamic World of Mobile Game Business', 1 April 2005, www.soberit.hut.fi/T-76.640/Slides/T-76.640_Rovio2005_04_01HUT.pdf.

7 'The 30 Best Pieces of Advice for Entrepreneurs', article on FirstRound.com, firstround.com/article/30-Best-Pieces#ixzz2pRF5EZ8a.

8 Ibid.

9 'Drew Houston's Morph from Hacker to Hyper-Growth CEO', op. cit.

10 Ibid.

11 'The 30 Best Pieces of Advice for Entrepreneurs', op. cit.

12 Ibid.

13 Eric Savitz, 'Jack Dorsey: Leadership Secrets of Twitter and Square', article for Forbes, 5 November 2012 issue, www.forbes.com/sites/ericsavitz/2012/10/17/jack-dorsey-the-leadership-secrets-of-twitter-and-square/.

14 'The 30 Best Pieces of Advice for Entrepreneurs', op. cit.

Chapter 39: Getting Acquired

1 Pitchbook, US, 'VC Valuations and Trends', 2014 annual report.

2 Wade Roush, 'Google's Rules of Acquisition: How to Be an Android, Not an Aardvark', article on Xconomy.com, 3 May 2012, www.xconomy.com/san-francisco/2012/03/05/googles-rules-of-acquisition-how-to-be-an-android-not-an-aardvark/?single_page=true.

3 Ibid.

4 J. O'Dell, 'Want to Get Acquired by Google? Google VP Explains How to Go Big', article on VentureBeat.com, 19 April 2012, venturebeat.com/2012/04/19/want-to-get-acquired-by-google-google-vp-explains-how-to-go-big/.

5 Mike Isaac, 'Facebook Acquisition Talks With Waze Fall Apart', article for AllThingsD.com, 29 May 2013, allthingsd.com/20130529/facebook-acquisition-talks-with-waze-fall-apart/.

6 Simone Wilson, 'Billion-dollar Waze', article on JewishJournal.com, 19 June 2013, www.jewishjournal.com/cover_story/article/billion_dollar_ waze.

7 Ibid.

8 For details of the merger see www.sec.gov/Archives/edgar/ data/1326801/000132680114000010/form8k_2192014.htm.

9 Gerry Shih and Sarah McBride, 'Facebook to buy WhatsApp for $19 Billion in Deal Shocker', article on Reuters.com, 20 February 2014, www.reuters. com/article/2014/02/20/us-whatsapp-facebook-idUSBR EA1I26B20140220.

10 'Facebook to Acquire WhatsApp', press release on FB.com, 19 February 2014, newsroom.fb.com/News/805/Facebook-to-Acquire-WhatsApp.

11 Justin Lafferty, 'Facebook Revenue Hits $2B in Q3, Now Has 507m Mobile DAUs', article on InsideFacebook.com, 30 October 2013, www.inside-facebook.com/2013/10/30/facebook-revenue-hits-2b-in-q3-now-has-507m-mobile-daus/.

12 So WhatsApp seems to be processing more messages than all SMS in the world.

13 'Facebook to Acquire WhatsApp', 19 February 2014, op. cit.

14 Adario Strange, 'Facebook Reportedly Offered $1 Billion to Acquire Snapchat', article on Mashable.com, mashable.com/2013/10/26/facebook-snapchat/.

15 Evelyn M. Rusli and Douglas MacMillan, 'Snapchat Spurned $3 Billion Acquisitions Offer from Facebook', blog post on WSJ.com, 13 November 2013, blogs.wsj.com/digits/2013/11/13/snapchat-spurned-3-billion-acquisition-offer-from-facebook/.

16 Cheryl Conner, 'Facebook's Reality Check: Death by a Thousand Snapchats?', article on Forbes.com, 8 July 2013, www.forbes. com/sites/cheryl

snappconner/2013/06/08/facebooks-reality-check-death-by-a-thousand-snapchats/.

17 Kara Swisher, 'Yahoo Tumblrs for Cool: Board Approves $1.1 Billion Deal as Expected', article on AllThingsD.com, 19 May 2013, allthingsd.com/20130519/yahoo-tumblrs-for-cool-board-approves-1-1-billion-deal/.

18 Todd Wasserman, 'Tumblr's Mobile Traffic May Overtake Desktop Traffic This Year', article on Mashable.com, 21 February 2013, mashable.com/2013/02/21/tumblr-mobile-traffic/.

19 Ibid.

20 Marc Andreessen, 'Why Software Is Eating the World', article on WSJ.com, 20 August 2011, online.wsj.com/news/articles/ SB10001424053111903480 904576512250915629460.

21 Leena Rao, 'As Software Eats the World, Non-Tech Corporations Are Eating Startups', article on TechCrunch.com, 14 December 2013, TechCrunch.com/2013/12/14/as-software-eats-the-world-non-tech-corporations-are-eating-startups/.

22 Alexia Tsotsis, 'Monsanto Buys Weather Big Data Company Climate Corporation for Around $1.1B', article on TechCrunch.com, 2 October 2013, TechCrunch.com/2013/10/02/monsanto-acquires-weather-big-data-company-climate-corporation-for-930m/.

23 Leena Rao, 14 December 2013, op. cit.

24 Ibid.

25 Pitchbook, US, 'VC Valuations and Trends', 2014 annual report.

26 'Yesterday's Big Payday for the IRS: 1600 Twitter Employees Now Millionaires', research on PrivCo.com, 8 November 2013, www.privco.com/the-twitter-mafia-and-yesterdays-big-irs-payday.

27 Sven Grundberg, '"Candy Crush Saga" Maker Files for an IPO', article on WSJ.com, 18 February 2014, online.wsj.com/news/articles/ SB10001424052702304675504579390580161044024.

28 'UK Mobile Games Maker King Delays IPO Due to Candy Crush Surge', article on VCPost.com, 9 December 2013, www.vcpost.com/articles/19437/20131209/uk-mobile-games-maker-king-delays-ipo-due-candy-crush.htm.

29 Phillipa Leighton-Jones, 'Why Candy Crush Is a Success That's Hard to Copy', blog post on WSJ.com, 18 February 2014, blogs.wsj.com/money-beat/2014/02/18/why-candy-crush-is-a-success-that-cannot-be-copied/.

30 Mark Berniker and Josh Lipton, 'Uber CEO Kalanick: No Plans To Go Public Right Now', article on CNBC.com, 6 November 2013, www.cnbc.com/id/101175342.

Index

Note: page numbers in **bold** refer to illustrations, page numbers in *italics* refer to information contained in tables.

About the Author

George Berkowski is a serial entrepreneur who has built businesses in manned space flight, online dating, transportation and mobile apps. He is one of the minds behind the internationally successful taxi hailing app Hailo, where he led the product team until September 2013. George studied rocket science and economics at the Massachusetts Institute of Technology, and business at the École Supérieure de Commerce de Paris. He was Chairman of MIT's Enterprise Forum in the UK and is heavily involved in the UK and US startup scenes. George now divides his time between London and Silicon Valley.

To find out more about George and keep up to date with the ever-changing world of billion-dollar apps, visit **www.mybilliondollarapp.com**.

You can also join the conversation on Twitter by following **@georgeberkowski** and using **#BILLIONDOLLARAPP**.